Randolph,

Dogface soldiers
moonlighting as authors --
who knew?

Dogface
Charlie

Here are some of
the stories as remembered
four decades later,
Best,
Patrick "Mac" McLaughlin

Cantigny Military History Series

Paul H. Herbert, General Editor
Steven Hawkins, Managing Editor

The Cantigny Military History Series is an activity of the First Division Museum, a part of the Robert R. McCormick Foundation, Chicago, Illinois. The purpose of the series is to publish materials that preserve the history of the U.S. Army's 1st Infantry Division and promote American military history and affairs. The First Division Museum is located at Cantigny Park, the historic estate of the late Robert R. McCormick, in Wheaton, Illinois, about thirty-five miles west of Chicago. For further information, contact the First Division Museum, 1 S 151 Winfield Road, Wheaton, IL 60189, tel. (630)-260-8185, or visit www.firstdivisionmuseum.org.

Other Books in the Cantigny Military History Series

Troubleshooting All the Way
A Memoir of the 1st Signal Company and Combat Telephone Communications in the 1st Infantry Division, 1944–1945
By Lovern "Jerry" Nauss

First to Warn
My Combat Experiences in the 1st Reconnaissance Troop, 1st Infantry Division, in North Africa and Sicily in World War II
By George J. Koch

A Century of Valor
The First Hundred Years of the Twenty-Eighth United States Infantry Regiment – Black Lions
By Col. Stephen L. Bowman, USA (Ret.)

The Beast Was Out There
The 28th Infantry Black Lions and the Battle of Ông Thanh
Vietnam, October 1967
By James E. Shelton

Blood and Sacrifice
The History of the 16th Infantry Regiment
From the Civil War Through the Gulf War
By Steven E. Clay

The Greatest Thing We Have Ever Attempted
Historical Perspectives on the Normandy Campaign
Edited by Steven Weingartner

Blue Spaders
The 26th Infantry Regiment, 1917-1967

No Mission Too Difficult
Old Buddies of the 1st Infantry Division Tell All About World War II
By Blythe Foote Finke

One More Hill
The Big Red One from North Africa to Normandy
By Franklyn A. Johnson

The Big Red One
America's Legendary 1st Infantry Division
from World War I to Desert Storm
By James Scott Wheeler
Copublished with University Press of Kansas

"Happy Days!"
By Alban B. Butler, Jr
Copublished with Osprey Publishing

Cantigny Military History Series

Dogface Charlie

Soldiers' Recollections of Vietnam and the Big Red One

Compiled by Tom Mercer

First Division Museum at Cantigny Park
Wheaton, Illinois
Part of the Robert R. McCormick Foundations

Dogface Charlie

Printed in the United States of America

ISBN: 978-1-890093-26-6
Library of Congress Control Number:

Design and Production: Wildenradt Design Associates - Evanston, IL

Photographs in this book were provided by Lauren Coleman, Kenny Gardellis, Darleen Keithly, Tony Kulikowski, Phil McClure, Patrick McLaughlin, Tom Mercer, Tom Pippin, Michael Shapiro, and David Sorensen.

Maps in this book were adapted and created by Ed Docekal.

Back cover: Swamp Rats and 18th Infantry Regiment patches from the Robert Mueller Collection, First Division Museum. 1st Infantry Division patch from the collection of the First Division Museum.

Dedication

This work is dedicated to the men who fought in the Vietnam War and their loved ones who waited for them to come home.

Some gave all; all gave some.

Welcome Home!

Dogface Charlie

Acknowledgements

The Dogface soldiers of Charlie Company extend our most sincere gratitude to Dr. Paul Herbert (Colonel USA, Ret.), Executive Director of the First Division Museum at Cantigny. We extend our heartiest thanks to his staff who worked on this book and especially to Steve Hawkins, Director of Publications, for his steadfast diligence in the editing and organization of the book. The First Division Museum at Cantigny made a commitment to edit and publish this book, sight unseen, and we will always be grateful for the confidence shown in a bunch of old combat soldiers moonlighting as authors.

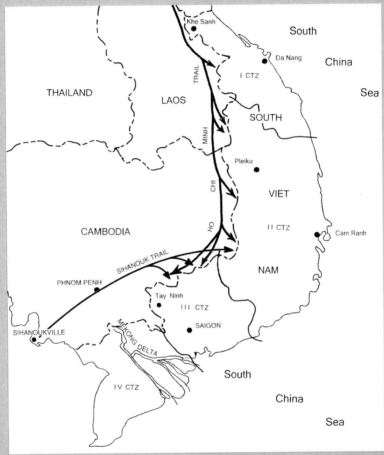

Corps Tactical Zone Boundaries and Enemy Trails

Table of Contents

Contents

Contents

1/18th Timeline in Vietnam

1965

July 12
USNS Gordon, Cam Ranh Bay. [Landing.]

August 11 – September 8
Operation BARACUDA, 12 miles west of Nha Trang, near Qui Nhon, 1/18 Inf, 2 Bde.

September 23
Bien Hoa. [1/18 Inf, 2 Bde. Base Camp.]

October 4-25
Operation HOPSCOTCH, Phouc Vinh, 1/18 Inf, 2 Bde.

November 1 – December 8
Operation VIPER, 1/18 Inf, 2 Bde. [Bien Hoa.]

December 17-22
Operation SMASH, 1/18 Inf, 2 Bde, 1st Inf Div. [Bien Hoa.]

1966

January 28 – February 15
Operation MALLET & MALLET II, Clear Highway 15 to Vung Tau, 1/18 Inf, 2 Bde.

February 21-27
Operation MASTIFF, Dau Tieng, 1/18 Inf, 2 Bde, and 3d Bde.

March 30 – April 15
Operation ABILENE, 1/18 Inf, 1st Inf Div.

April 18
Bear Cat [1/18 Inf, Base Camp].

April 24 – May 17
Operation BIRMINGHAM, 1/18 Inf, 1st Inf Div. [Di An]

July 9
Minh Thanh Road, 1/18 Inf.

May 21 – June 9
Operation LEXINGTON, Rung Sat Special Zone, 1/18 Inf.

July 13 – August 24
Operation EL PASO III, An Loc, Quan Loi, Loc Ninh, Minh Thanh, patrols 1/18 Inf.

September 4 – October 8
Operation BATON ROUGE, Rung Sat Special Zone, 1/18 Inf, 2 Bde.

October 21-28
Operation LAM SON [aka ALLENTOWN], SW of Di An, Saigon River, 1/18 Inf, 2 Bde.

November 5-25
Operation ATTLEBORO, Tay Ninh Province, 1/18 Inf, 1st Inf Div.

November 29
Operation LAM SON, [Civic Action and patrols] Phu Loi, 1/18 Inf.

1967

January 8-26
Operation CEDAR FALLS. 1st Inf, Div.

January 28
End Operation LAM SON, [Civic Action and patrols] Phu Loi, 1/18 Inf.

February 22 – April 15
Operation JUNCTION CITY I & II, Quan Loi, Highway 13, and Minh Thanh, 1/18 Inf, 1st Inf Div.

April 23 – May 11
Operation MANHATTAN, 1/18 Inf, 1st Inf Div.

May 17-25
Operation DALLAS, Tan Uyen, 1/18 Inf, 1st Inf Div.

July 29
Village of Cha Tieu, 1/18 Inf, 2 Bde.

July 29
Operation ROCHESTER, [Lai Khe], [rocket attack, on 2 Bde.]

September 29 – November 19
Operation SHENANDOAH II, Chon Thanh, 1/18 Inf, 1st Inf Div.

October 6
Da Yeu, B, C, D Cos, 1/18 Inf.

October 29
Srok Silamlite I, C, D Cos, 1/18 Inf. engaged the 165th NVA. (CIDG Co also pursued the enemy. A Co manned the NDP.)

October 30
Srok Silamlite II, A, D Cos, 1/18 Inf. (Bn Cmd Group with CIDG Co. C Co manned the NDP.)

November 2
Srok Silamlite III, A, C, D Cos, 1/18 Inf.

December 8
Xa Cat, A Co, 1/18 Inf.

1968

January 31 – February 20
Enemy Tet Offensive, Tan Son Nhut Air Base, Saigon, 1/18 Inf.

January 31
On the Saigon River delta Company A, from Camp Diamond on Song Tac River 5 miles east of Thu Duc, made their first amphibious assault since WWII against a VC base camp on the Song Dong Nai River.

January 31
Near Thu Duc Companies A and B air assaulted to help defend the ARVN subsector headquarters against a VC battalion.

February 5 (or 7)
South of the Thu Duc water purification plant, Companies B and D, a platoon of Company A, and the Reconnaissance Platoon engaged an enemy company at an oil refinery.

February 8
Company C defended the vital Thu Duc thermo-electric plant and US AID rice warehouses.

May 5
Xom Moi II, 2nd Platoon, B Co, 1/18 Inf.

October
Lai Khe, Phouc Vinh, 1/18 Inf.

November
Di An, Catcher's Mitt, 1/18 Inf, 2 Bde.

1969

January 1-11
Operation TOAN THANG II, 1/18 Inf, with the 11th Armored Cav. [Bien Hoa]

January 12 – March 11
Operation TOAN THANG II, Lai Khe [prevent rockets], 1/18 Inf.

September 15-26
Phu Hoa Dong Seal, A, B Cos, 1/18 Inf.

September 23
Fire Support Base Normandy III, Catcher's Mitt, 1/18 Inf.

November 10-11
Ambush in Catchers Mitt, A Co, 1/18 Inf.

List of Maps

Terms Used in Vietnam

11 ACR
11th Armored Cavalry Regiment, called "Black Horse" for their emblem

11B
Military Occupation Specialty (MOS) designation for light infantry

ACR
Armored Cavalry Regiment

AIT
Advanced Individual Training

AK-47
Russian fully-automatic assault rifle

AO
Area of operations

APC
Armored personnel carrier

ARTY
Slang for artillery

ARVN
Army of the Republic of Vietnam

Beehive Round
Artillery round containing thousands of small projectiles called "flichettes"

Bien Hoa
Both a large Air Force base and city about 20 miles northeast of Saigon

Berm
Fortification or emplacement

BN
Battalion

Boonies
Slang for the field

Bush
Slang for the field

C-4
Plastic, putty-like explosive carried by infantrymen

Cache
Hidden supplies

CAV
Cavalry

CHI-COM
Chinese Communist

Chieu Hoi
Surrender

Chinook
CH-47 medium lift helicopter

CIB
Combat Infantry Badge

CIDG
Civilian Irregular Defense Group

Claymore Mine
Command-detonated antipersonnel mine, designated M-18A1

Clicks
Kilometers

CO
Commanding Officer

CP
Command Post

C-Rations or C-Rats
Combat rations; canned meals for use in the field

DEROS
Date of estimated return from overseas

Deuce-and-a-Half
Two-and-a-half ton truck; standard military troop and supply truck

Di An
Base headquarters for 1st Infantry Division, 2nd Brigade, and 1-18th Inf after Sept. 1966 1/18 Base Camp

Dust-off
Aerial evacuation of the wounded and KIAs

Flanker
The person who walks to the side of the patrol

FB
Fire base. A temporary artillery encampment providing fire support to forward operations

Firefight
Exchange of small-arms fire with the enemy

FNG
F°°°°°g new guy

FO
Forward observer for artillery

FRAG
Fragmentation grenade, typically designated M-61

Gooks
Slang term for VC/NVA

Gunship
Heavily armed helicopter used to provide fire support for infantry

H AND I
Harassment and interdiction fire.

HE
High explosive

Hooch
Makeshift shelter

Hot or Hot LZ
Area under fire, as in hot landing zone (LZ)

Huey
UH-1 Iroquois helicopter

KIA
Killed in action

KP
Kitchen police. Pulling duty
in a mess hall

Lai Khe
Base headquarters of the 1st
Infantry Division in 1968

LAW
Light anti-tank weapon, shoulder
fired 66mm rocket launcher

Leg
Slang term for infantryman

LOH
Light observation helicopter,
typically an OH-6A "Cayuse"

LP
Listening post, a 2-3 man
position set up at night outside
the perimeter or berm

LRRP
Long-range recon patrol

LT
Lieutenant

LZ
Landing zone

M-2
.50 cal machine gun

M-14
Early U.S. standard infantry
rifles, a 7.62mm (.308 caliber)
rifle with a 20 round magazine
that replaced the M-1

M-16
Standard U.S. infantry weapon
in Vietnam; 5.56mm semi-auto
and auto rifle with a 20 round
magazine

M-60
Standard light-weight 7.62mm
(.308 caliber) machine gun

M-61
Standard-issue fragmentation
grenade for infantry soldiers

M-79
Single-shot 40mm grenade
launcher. Also referred to as a
"blooper" or "thump gun" due
to the sound it made when fired.
Could fire HE canister (buckshot)
or "flechette" rounds

MAC-V
Military Assistance Command,
Vietnam

Mamasan
Slang for older Vietnamese
female

MEDEVAC
Medical evacuation

MIA
Missing in action

Mortar
Small, indirect-fire weapon for
infantry. Typically 81mm or 4.2in

MSR
Main supply route

NCO
Noncommissioned officer

NCOIC
Noncommissioned officer in charge

NDP
Night defensive position or perimeter

NVA
North Vietnamese Army; regular soldiers as opposed to Viet Cong guerrilas

OP
Observation post, typically referring to daylight positions

Oscar Platoon
Mortar platoon

Papasan
Slang for older Vietnamese male

Patrol
Small group of infantrymen

Perimeter
Outer defensive limits of a military position

PFC
Private first class

Platoon Leader
Usually a 2nd Lieutenant, typically leading 3-4 squads of infantrymen

Point Man
Forward or first man on combat patrol

PRC-25
Portable radio Model 25. Standard radio carried by platoon-level or smaller infantry units. Frequencies were changed by changing the crystal in the radio.

Quarter Cav
1st Squadron 4th Cavalry, assigned to the 1st Infantry Division

R & R
Rest and recreation a 3-7 day leave from the war zone

Rear Security
Last on a patrol or in a line of units

RECON
Reconnaissance

REMF
Rear echelon motherf°°°ers

RIF
Reconnaissance in force

RON
Remain-overnight

RPG
Rocket-propelled grenade

RTO
Radio telephone operator, usually carried the PRC-25 in squad or platoon formations

SIT-REP
Situation report

Slick
Slang for the Bell UH-1 Iroquois helicopter other than a gunship or medical evacuation aircraft

Smoke Grenade
Used to signal locations. Came in colors (yellow, green, blue, and red for emergencies only)

SOP
Standard operating procedure

SP
Specialist

SSG
Staff sergeant (E-6), typically a squad leader

Starlight Scope
Night-vision optical telescope

TET
Buddhist Lunar New Year

Thump Gun
M-79 grenade launcher

TO&E
Table of organization and equipment

TOC
Tactical operations center

Trip Flare
Ground flare triggered either by a tripwire or by a spring-loaded handle

UH-1
Bell UH-1 Iroquois helicopter (Huey), the primary infantry troop mover and supply aircraft

VC
Viet Cong

Victor Charles/Charlie
Viet Cong

WIA
Wounded in action

WP Willy Pete
White phosphorous grenades

XO
Executive officer, second in command

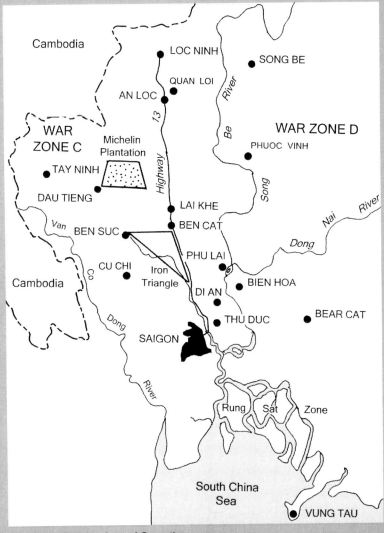

1st Infantry Division Area of Operation

Preface

Paul Herbert, Ph.D.
Colonel, U.S. Army (Retired)

This book is by and about some of the soldiers who served in Vietnam as members of Company C, 1st Battalion, 18th Infantry, 1st Infantry Division, especially in 1967 and early 1968. It is not a history of the policies that led to the American intervention in Vietnam nor is it a history of the strategies and military campaigns undertaken there. Instead, it describes these soldiers' experiences as remembered some forty years afterward. Their stories underscore a central fact of that unpopular war – the United States sent very good soldiers to Vietnam and they fought very well.

For the 1st Infantry Division, the famed "Big Red One," the war in Vietnam unfolded in four phases between April 1965 and April 1970. The first phase was the Division's deployment from Fort Riley, Kansas, its establishment in its area of operations north of Saigon in the III Corps Tactical Zone (CTZ) between June 1965 (when the 2nd Brigade, including Charlie Company, left Fort Riley) to March 1966, and the assumption of division command by Major General William E. Depuy. The second phase consisted of "search and destroy" operations by large formations throughout III CTZ and lasted through the 1968 Tet Offensive in January and February of that year. The third phase of strategic transition grew directly out of the Tet Offensive and lasted another year, until President Richard M. Nixon's inauguration in January 1969. It included the decisions in March by President Lyndon B. Johnson not to send significant additional forces to Vietnam and not to seek another term as president; the change of theater command in Vietnam from General

William C. Westmoreland to General Creighton W. Abrams, in July; and the initiation of a policy of pacification and "Vietnamization" under Abrams and formalized by the new Nixon Administration in 1969. The final phase, Vietnamization, included all of 1969 and ended with the Division's return to Fort Riley between January and April 1970.

Throughout the war, the 1st Infantry Division operated in the III CTZ, one of four such military zones organized by the South Vietnamese to protect four key strategic areas: the De-militarized Zone (DMZ) boundary between North and South Vietnam in I CTZ in the north; the narrow "waist" of Vietnam in the Central Highlands between Cambodia and the South China Sea in II CTZ; Saigon, the capital, in III CTZ; and, furthest south, the rice-rich Mekong River delta in IV CTZ.

The essential military problem in III CTZ was to protect Saigon and the indigenous population from Viet Cong main force units whose bases were in the thick tropical forests along the north-south rivers such as the Saigon River and also in the French-owned rubber plantations in the interior. These units were supported (and eventually replaced) by the North Vietnamese Army from bases in supposedly neutral Cambodia, the southern terminus of the famous "Ho Chi Minh Trail" infiltration route from North Vietnam. The serpentine Cambodian-Vietnamese border in III CTZ meant that some enemy bases in Cambodia were less than 50 miles from Saigon. The American presence in III CTZ came under a corps-sized headquarters in Long Binh, north of Saigon, called II Field Force. It eventually included not only the 1st Infantry Division but also the 25th and 9th Infantry Divisions; the196th and 199th Light Infantry Brigades; the 173d Airborne Brigade; the 11th Armored Cavalry Regiment; and the Royal Australian Regiment, along with a host of supporting formations.

Charlie Company arrived in Vietnam with the 2nd Brigade, 1st

Infantry Division, aboard the United States Naval Ship (USNS) *Gordon* on July 12, 1965, after a three-week voyage across the Pacific from San Francisco. Initially, Charlie Company and the rest of the 1st Battalion, 18th Infantry (1-18 Infantry) helped secure the critical logistical port of Cam Ranh Bay in II CTZ, but by September had rejoined the 2nd Brigade at its new base camp at Bien Hoa just north of Saigon.

The 1st Infantry Division area of operations (AO) for most of the war was a V-shape centered on National Highway 13 leading due north from Saigon some 90 miles through Binh Duong and Binh Long Provinces to the Cambodian border. The AO was bounded very roughly on the west by the Saigon River, its western half called by its old French designation, War Zone C; and on the east by the Song Be River, its eastern half called War Zone D. The AO included the "Iron Triangle" Viet Cong base area just north of Saigon; the Michelin rubber plantation further north; and the important commercial and political centers of An Loc and Loc Ninh, also amid large rubber plantations. The division headquarters set up initially in Di An but later moved north to Lai Khe. The Division's mission was to search out and destroy enemy units infiltrating through its AO from Cambodia to Saigon while protecting the population and extending the credibility and effectiveness of the South Vietnamese government.

This same region was known to the enemy as the "B2 Front." Beginning in 1965, it was home to the 9th Division of the Peoples Liberation Armed Forces (PLAF), a regular Viet Cong (VC) formation consisting of three infantry regiments and other supporting units. These were augmented and aided as needed by local VC guerrillas and often reinforced by North Vietnamese Army (NVA) cadres and line regiments operating out of bases in Cambodia.

1st Division units went immediately into the fight. For the 1-18 Infantry, this meant securing a base camp area near Phuoc Vinh for the arriving 1st Brigade, 1st Infantry Division; clearing Highway 15 as a convoy route from Bien Hoa east and south of Saigon to the port of Vung Tau; and conducting airmobile operations as far north as An Loc and west as Dau Tieng. Highway 15 passed just east of the "Rung Sat Zone," a mangrove swamp delta of the Dong Nai River southeast of Saigon believed to be a major staging area for the VC. In Operations LEXINGTON (May and June, 1966) and BATON ROUGE (September and October, 1966), the 1-18 Infantry was committed to clearing the Rung Sat Zone, a dubious proposition for a unit lacking the boats, special equipment and training needed for continuous operations in swampland. The soldiers made little contact but did uncover enemy facilities and supplies and earned the battalion the informal nickname "Swamp Rats," a term that stuck with them for the rest of the war. Still later, they became "Dogfaces," after the radio code name for the battalion, which was Dogface.

By early 1967, the Swamp Rats were an experienced, veteran unit, perfectly attuned to the "search and destroy" strategy of General Westmoreland and the airmobile doctrine by which that strategy was carried out, especially by General Depuy. Because the ground war could not be carried into Cambodia, Laos, and North Vietnam, Westmoreland reasoned that it had to be won within South Vietnam. US forces would be used against the most dangerous enemy forces, the "main force" VC units and the NVA. These had to be detected and destroyed at a rate that exceeded the North Vietnamese ability to replace them by infiltration. With enemy regular forces so engaged, the South Vietnamese government would have breathing space for the internal security and civil development that would garner the loyalty of the people and win the war politically.

To find, engage, and destroy enemy forces at the least cost in US casualties in the vast tropical expanses of Vietnam, US forces would exploit mobility, especially the relatively new assault heli-

1968. Landing in an LZ Outside of Lai Khe

copter, and the mobile firepower of artillery, helicopters, naval gun-fire, and Air Force, Navy, and Marine fighters and bombers. General Depuy, who had been Westmoreland's J-3 or Assistant Chief of Staff for Operations, developed appropriate tactical procedures as Commanding General of the Big Red One. Helicopters would insert infantry into suspicious areas by surprise. Infantry used "clover-leaf" patrols to find the enemy with the smallest force possible. More infantry would "pile on" in multiple helicopter insertions to box the enemy in. All available fires, pre-arranged and immediate-ly available, were called in to destroy him. Infantry then maneuvered to destroy enemy base camps and supplies, or assumed a thorough-ly prepared "night defensive perimeter" (NDP) to tempt remaining enemy forces into attacks that rendered them vulnerable to over-whelming US firepower. Depuy specified such tactical techniques right down to the configuration of soldiers' fighting positions.

Throughout 1966 and the first half of 1967, when command passed to Major General John H. Hay, the Big Red One pursued this strategy relentlessly.

Major operations included CEDAR FALLS (January 1967), JUNCTION CITY (February to May), MANHATTAN (April to May) and BILLINGS (June). Each of these was designed to trap large enemy forces in their presumed base areas. CEDAR FALLS was directed against the "Iron Triangle" closest to Saigon as a prelude to the main effort, JUNCTION CITY, much farther north in Tay Ninh Province, where commanders believed the enemy headquarters called the Central Office for South Vietnam (COSVN) was located. JUNCTION CITY involved nearly all the combat forces in II Field Force, including a parachute assault by the 173d Airborne Brigade. MANHATTAN concentrated on the so-called Long Nguyen Secret Zone, a transit area between War Zones C and D just east of the Michelin rubber plantation. BILLINGS was a foray into War Zone D. The upshot of all these operations was that few enemy forces were destroyed relative to the allied forces engaged. Casualties were inflicted, to be sure. Supplies and facilities were destroyed and enemy operations were seriously disrupted. However, major enemy formations and headquarters were able to avoid contact with US forces and either go to ground (sometimes literally, in extensive tunnel networks) or withdraw into Cambodia.

By the summer and early fall of 1967, the situation on the battlefields of Vietnam led allied and North Vietnamese leaders to complementary decisions. General Westmoreland believed that enemy casualties were approaching the "cross over point" at which they exceeded North Vietnamese infiltration rates – there was "light at the end of the tunnel" – and that vigorous pursuit of the enemy would buy the time needed to begin turning the war over to the Vietnamese. In III CTZ, this meant, among other things, opening Highway 13 all the way to Loc Ninh and keeping it secure to stimulate commerce between Loc Ninh and Saigon. North Vietnamese leaders believed that the casualties they suffered that summer required reconsideration of their long-war strategy. They determined to win the war in a single stroke early in 1968 with a general uprising throughout South Vietnam and especially in its coastal cities, including Saigon. As a prelude to the uprising, US forces had to be drawn into the interior. To do that, bases and cities in the interior and along the DMZ would be threatened – Khe Sanh, Con Thien, Dak To – and, in III CTZ, Loc Ninh and Song Be near the Cambodian border.

These decisions precipitated Operation SHENANDOAH II that brought Charlie Company and the rest of the 1-18 Infantry to the area around Loc Ninh in October and November, 1967. This was some of the 1st Division's hardest fighting of the war, precipitating two Medal of Honor actions in the space of three weeks. The Americans were determined to find and finish off a reclusive enemy they had battered throughout the summer and to clear Loc Ninh and Route 13, the GIs' "Thunder Road," once and for all. The enemy, replenished and re-equipped from bases in Cambodia, was determined to threaten the US Special Forces camp and airstrip at Loc Ninh, as well as the provincial capital itself, in order to draw US attention and forces away from their infiltration into Saigon. Some of the most graphic accounts that follow are from 1-18 Infantry operations around Loc Ninh in the fall of 1967. When the fighting ended and 1-18 Infantry returned to its base camp at Quan Loi, the soldiers undoubtedly perceived a respite in the pace of operations. That was not to be – 80,000 Viet Cong troops struck at cities and bases throughout Vietnam on January 30 and 31st, 1968, during the traditional truce period of the Tet new year holiday. The 1-18 Infantry rushed by helicopter to help clear and secure Tan Son Nhut airbase in Saigon on January 31. Afterwards, they were assigned to secure the city water works, a mission that required

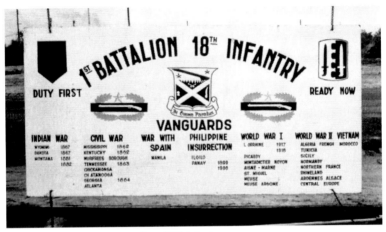

Di An. Home of the 1st Battalion, 18th Infantry.

operations on urban terrain significantly different from the tropical forests with which they were so familiar.

The Tet Offensive devastated VC formations throughout Vietnam but also sapped American will to continue the war. After Tet 1968, there were fewer major engagements for the 1-18 Infantry.

Operating out of Lai Khe and Phuoc Vinh, the battalion continued reconnaissance missions, air insertions, ambushes, and route clearance operations. It helped secure the firebases that stretched across the division's AO. Its contacts with the enemy, always deadly, were fewer and smaller. In April 1968, the term "search and destroy" began to disappear from unit after action reports and, beginning in July 1968, with General Abrams's succession to command, more emphasis was placed on the "Dong Tien" (Progress Together) program of operating with South Vietnamese counterpart units and on civic action to stabilize and protect the indigenous population. The battalion rejoined the 2nd Brigade at Di An in November. In 1969, Division commander Major General A.E. Milloy cited the relative security in III CTZ as a reason that the 1st Division was to be the first major combat unit withdrawn from Vietnam beginning in January, 1970.

Much of the literature of the Vietnam War concentrates on the highest political and military levels – the difficult intelligence, the competing strategies, the powerful personalities, the volatile politics, the wrenching decisions, the tragic consequences. Through all those long, difficult years of decision, some very good young men went to Vietnam and served and sacrificed nobly. This book recounts the stories of a few – the Swamp Rats and Dogfaces of Charlie Company, 1st of the 18th Infantry, soldiers of the Big Red One.

Paul H. Herbert, Ph.D.
Colonel, U.S. Army (Retired)
General Editor, Cantigny Military History Series
Wheaton, Illinois
January 2012

Introduction

George M. Tronsrue, Jr.,
Colonel, U.S. Army (Retired)

This book began as an idea in the head of one of our riflemen, a line combat soldier – although now quite a few years removed from that role. Tom Mercer was a fire-team leader in Company C, one of the four rifle companies in the 1st Battalion, 18th Infantry as it was organized for the war in South Vietnam in 1967-68. The battalion was best known by its radio call sign at the time: Dogface Battalion. It was at Tom's prompting and through his actions that this book on Dogface began to take shape. More than 20 other members of the battalion, sensing that the book's concept could lead to uncharted paths, responded to his urging. In their own ways they produced these individual descriptions of a unique time in our military history. Keep that fact in mind as you delve into these pages. These are the writings of individual infantrymen telling about a traumatic time in their lives. None of these writings flowed from the pens of professional military authors. All of them began, and we hope still are appreciated, as the thoughts and ideas of everyday soldiers, proud of what they and their buddies did in the mid-1960s in South Vietnam.

The stories of the Army's rifle companies that served in South Vietnam are unique as to people, places, and what they actually did. The experiences, especially, of the members of rifle companies are as different as the people describing them.

My own thoughts about Vietnam, for example, have many dimensions. My experience began in 1962-63 when, as a senior captain, I was assigned as one of two liaison officers to the Military

Loc Ninh. Oct 29-Nov 5, 1967, CIDG Unit (possibly ARVN).

Assistance Command, Vietnam (MAC-V). My duty was as an observer with ARVN units at the battalion and regimental levels in I Corps in the northern parts of the country. This was followed directly by duty in the Mekong Delta as the initial battalion advisor to the 17th Civil Guard Battalion. The battalion commander reported directly to the province chief of Dinh Tuong Province. (I accompanied individual companies of the battalion on some 23 company-level combat operations in 1962-63.) Finally, after that introduction, I served in 1967-68 as the commander of Dogface Battalion.

Having had that fairly broad exposure to operations at the company and battalion levels, when I first saw Charlie Company, I was not learning so much as I was comparing with past experiences just how well our 18th Infantry companies faced daily combat in 1967-68. And I was humbly proud to play a part in their success.

In the pages to follow, you will read first-hand experiences of our young men in combat in the mid-1960s in South Vietnam. I hope that you will react with a "Wow!" in some cases, a tear or two in others, and a general feeling of admiration for just how well our young men handled a war so different from all that preceded it.

George M. Tronsrue, Jr.,
Colonel, U.S. Army (Retired)
Battalion Commander,
1st Battalion, 18th Infantry Regiment,
1967-68 (Dogface Six)

3rd Squad, Mike pit, from left: Sp4 Russles Simms, Sp4 Ed (Butch) Johnson, Sp4 Tom Murphy, SGT Robert Shand, Sp4 George Elmore.

Duty and Comradeship

Richard E. Cavazos, General, U.S. Army (Retired)
Battalion Commander, 1st Battalion,
18th Infantry Regiment, 1967 (Dogface Six)

The Dogfaces measured up to every challenge. We lost good men that grieve us to this day, but loyalty, pride, and comradeship come out in large measure. Young men who never dreamed they would receive the call to duty and country responded well beyond what could have been expected. These young heroes, yes first-class soldiers and heroes, responded well beyond what should only be expected of military professionals.

The battles we fought together were frequent and brutal. Loyalty, reliance, and trust in one another were fostered – duty was more than just a byword. Perhaps we all grew into soldiers and the call to duty was much more. The code of duty, honor, and country was real and remains enduring to this date.

While we punished the enemy, at no time were the lives of Dogface soldiers valued less than the battlefield objectives we pursued. Therein lies the superb blend of duty and comradeship.

Yes, we were honored by a Valorous Unit Award, but long before the award I knew that Charlie Company – that all of Dogface – was an exceptional group of soldiers who looked after each other and punished our enemies.

Somewhere the idea of duty, of love for our land, and comradeship grew paramount and blossomed into a unique fighting force that other soldiers in the 1st Division envied, respected, and admired. You were truly superb fighters, loyal to the colors and to each other. All of you know you were good combat infantrymen.

How good, you never suspected until much later.

Bless you for the soldiers you still are. I consider it one of the great honors of my military career to have commanded, fought, and soldiered with you.

Swamp Rats Story

Larry Van Kuran
Yep, boats it was.

The 1-18th Infantry first went into the Rung Sat Swamp in April 1966, before I was there. Last time was September-October 1966. I made that run. Rhun Sat was a tidal (12-hour) salt-water mangrove swamp. Very, very hard to stay dry and out of the salt water, which does things to your body that you can't believe. Not like fresh water at all. We used to buy those plastic hammocks from the kids in town, the ones with the finger-thick white plastic rope. We tied the hammocks onto mangrove trees, so that we could stay up out of the salt water when we slept. Those who didn't have hammocks would have to try to sleep on the mangrove tree roots – very difficult to do.

I remember the very first time I went in the Rhun Sat. We went up the Saigon River in Higgins Boats – the WWII landing craft with the drop-down front ramp. Some of us were commenting that they were trying to make us into John Wayne types and storm the beach like Normandy (Longest Day movie).

We went up the river at low tide. Dropped the ramp in the mud. The Navy guys running the boats showed us how we had to get up the mud bank. Because the mud was so deep, you couldn't stand up; all that kept you from sinking down up to your chin was your crotch. And the mud was so thick that it would suck your boots right off your feet. So, we had to lie on our stomachs and kind of "crab" our way up, wiggling back and forth, side to side. When we got to the top of the bank, we had a mud pack about 3-4 inches thick that we had to scrape off our chests and legs.

We'd go in and stay in the swamp 2-3 days, then come out to dry out at Vung Tau for 2-3 days, then back in again. You know how we knew we were going back into the swamp? Well, the day before we would go back in, the kids would raise the price of their plastic hammocks to $5 MPC (military payment certificates); they usually sold them for 50 cents. They always knew before we did – probably from the Navy guys who had to prepare their boats to take us.

I remember the second time we went in, and the Higgins Boat dropped its ramp. About half the platoon was in the process of wiggling up the bank when we started taking fire from a single sniper from the bush behind the other river bank. I was about 2/3 of the way up, and a round slapped the mud about two feet to my right. Do you have any idea what it's like knowing that you are simply sitting there, waiting to get hit, with almost nothing you could do about it except try to wiggle a little faster? It was just about the most helpless feeling I had during my whole tour. It ended up that the guy was a real bad shot (or we were real, real lucky that day) – no WIA/KIA for us.

One time when we were in, I had developed some pretty bad ulcers on the tendons down the backs of my feet. John "Doc" Hollins, our medic, couldn't get them to heal because of the salt water in the swamp. Anyway, at the end of the 2nd day, they sent me back in just before dusk on a larger LCI(L) (Landing Craft Infantry [Large]). It was getting dark. The boat had some special weapons on it: a 4.2 inch mortar set into the floor and a twin 20mm near the pilothouse. One of the Navy guys asked us if we wanted to "test fire" it. Are you kidding? You bet. Never in my life had I felt so powerful as when I watched trees about 3-4 feet in diameter get literally chopped in half.

When we would come out, they'd use the boats a notch larger than the Higgins boats – I think they were LCI(L). They could hold an entire company. Anyway, the very last time the 1-18th Infantry came out of the Rhun Sat, they did a four-LCI convoy. We were the third boat in line. On the way down the river, out of the swamp, a VC detonated an underwater mine (used an underwater wire/cable to set it off). The mine went off under the fourth boat, right behind us. It blew three or four guys off the boat into the water. Several of us in the third boat started taking off our boots to go in after those guys. Gary Harbison went in and later received the Soldier's Medal. (The Soldier's Medal is awarded to any person of the Armed Forces of the United States or of a friendly foreign nation who, while serving in any capacity with the Army of the United States, distinguished himself or herself by heroism not involving actual conflict with an enemy. [Army Regulation 600-8-22]). Long story longer, we couldn't find the guys who got blown overboard. One of them was a black kid – found him about five days later, body so swollen and discolored from being in the water that you couldn't tell if he was black, white, or whatever.

During our stay in Vung Tau when we were about through going back in, they had us start training others to take our place – how to live in the swamp. And guess who we had to train? Marines. The good ol' USMC. The NCOs were all ears and really wanted to learn from us. They'd come over to where we were bunked in the evening just to talk more. But the basic USMC grunts? They wouldn't listen to Army guys. I wonder how many of those Jarheads ended up on the short end (i.e., having to sleep in the saltwater, etc.) from not having listened.

So, where did the Swamp Rat patch originate? I can tell you. One of our 1st Squad members got an idea for a patch and took a couple of the guys with him to a mamasan in a small village just off the Vung Tau airstrip. We went back a couple of days later and returned with 30 patches – I have one of those first 30 (very rare these days). About 30 years later some of us were looking at my patch. It was only then that we saw that mamasan had put her brand on it and screwed us. The rat is wearing a Chi-Com helmet with a red star, holding an AK-47, and sitting on an NVA flag. Gotta laugh. I've had patch collectors offer me $350 for that original patch. Every time I go to the reunions that we hold in conjunction with the Society of the First Division, the guys from the First Division Museum always come find me and ask me to donate the patch to them. Nope. Not yet.

Night Ambush Patrol

Larry Van Kuran

October 23, 1966. Out about 1-2 clicks we set up an ambush along rice paddy dikes with suspected VC traffic. The squad set up at dusk, lying partially in the rice paddy along the side of one of the dikes. Guys were half in/half out of the water. I was on Tom Murphy's right side, Kenny Quitt to my right; I can't recall the order of the others. Boy, if you don't think you can be cold when it's 90 degrees out?! We were shivering.

About 0300 or so we heard low voices and footsteps coming along one of the dikes toward us. It was dark, so we couldn't see until they got close – five or six VC carrying weapons. While we were lying there, and just as we spotted the VC coming down the dyke, a great big centipede started crawling up my leg (had a tear in my fatigue pants), and I was freaking out. Murphy laid his hand on my arm and said, "Van Kuran, not a sound – eat it!" I did. I reached down and pulled it out, but I still have dreams about that god***ed critter, and there were leeches, lots of them, all over.

Tom with the M-60 and the rest of the squad opened up at about 50 meters. We dropped four of the six; Tom got two or three (all KIA); either Kenny Quitt or I got number four (wounded). All four VC fell off the other side of the dike where we couldn't see them. The one Ken or I wounded called for help all night long. Every once in a while we'd toss a couple of grenades over to shut him up, but no luck. The moaning lasted the rest of the night until dawn (about 0330 to 0500). This was a long time to have to lay there and listen to it.

At dawn the squad did an end-around sweep toward the dike to check things out. The wounded VC was still alive – he rose up and tried to fire his .45 caliber pistol. Two of us finished him with M-16s. In case of booby traps, we tied lines to the legs of other VC dead and dragged them out of the paddy from a distance – no traps found.

When we left, we propped one of the VC up against a tree with a torn off 1st Infantry Division patch in his teeth as a social comment. We passed by the next afternoon (36 hours later), and he was still there, but the patch was gone. I guess we'd made our point; lines of ants were crawling up his neck.

The Buffalo Tale

Larry Van Kuran

Mid-December, 1966. Somewhere near Phu Loi, North of Saigon. Daylight patrol out of our base camp (we even had tents) with squad of 10 men, including (as I recall) Nathan

April 1968. Rice paddies.

Middleton (M-79), Tom Murphy (M-60), SGT "Choo Choo" Justice, myself (M-16), Larry Wilbourne (M-16), Ronald Wooten (M-16), Louis Quinones (M-16), David Lewis (M-16). We were carrying all our ammo; most guys had two to three M61 frags, a few had white phosphorous (WP) grenades, C-4 in two kilo sticks, yellow/green/red smoke, etc. Lots of firepower, though at squad level.

We had been out about two to three hours going through rice paddies. We came to another paddy and decided to cross diagonally from one corner instead of going the long way around. We never walked up on the top of the dikes.

First guys stepped over the dike. Then we noticed the farmer's "tractor" (his water buffalo) two paddies over. Buffalo are pretty protective and have the 1,000 lbs. to make that stick if they want. The buffalo saw us and put his ears down as they do when they are irritated, scared, or angry.

We kept stepping over the dike, one by one, and proceeding diagonally across the paddy, all the while watching the buffalo. Then we saw the buffalo place one foot in front of the next, and he hunkered down lower to the ground. As we proceeded farther across the paddy, the buffalo took more very slow steps in our direction, one at a time. By the time all ten of our squad had stepped over the dike, the buffalo was almost low-crawling and still stepping slowly toward us, his eyes now very wide and staring at us.

Sergeant Justice told us to pick up the pace. We did…and so did Mr. Buffalo.

We had made it just about halfway across when Mr. Buffalo put one foot across the dike and into the next rice paddy over from us. Remember, he had been about in the middle of a paddy two paddies over getting closer and closer.

We picked it up even more, as quickly as we could step in a wet paddy without running and splashing. And Mr. Buffalo kept stepping, one carefully placed foot at a time, but a bit more quickly now.

When we got about three quarters of the way to the opposite corner of our paddy, Mr. Buffalo made his first step across the dike into the same paddy we were in, eyes wide, still low but now twitching and the nose sniffing and puffing.

Then Mr. Buffalo started quick-stepping toward us. I remember "Choo Choo" Justice yelling at us not to shoot, to just move out smartly.

Here's a full 10-man infantry squad, sufficiently armed to be able to blow up a small village by ourselves. And here's this totally "unarmed" pet buffalo, all 1,000+ lbs. of him, coming toward us. And there was absolutely no doubt in anyone's mind as to what he intended to do – have himself some 11Bs (riflemen) for lunch.

Just as Mr. Buffalo started to do a real quick step in our direction, something amazing happened. Out of nowhere comes this kid about 8-10 years old, real, real small, about 70 lbs. I'd guess. The kid was carrying a switch – not a big stick, just a switch.

The kid runs in between Mr. Buffalo and us, closer to Mr. Buffalo. He gets to Mr. Buffalo, reaches out and grabs the nose ring and starts swatting Mr. Buffalo about the head and shoulders smartly with the little switch and yelling at him in his high-pitched Vietnamese voice of a kid.

And here's that amazing part: Mr. Buffalo – all 1,000+ lbs. of him – simply drops to his knees and cowers, trembling at every hit of the kid's switch. Smack, smack, smack. And Mr. Buffalo sank lower and lower into the water of the paddy with each smack.

So, here we are. The great big, fully armed-to-the-teeth herd of 11B tough guys literally starting to run for our lives from a family pet and farmer's "tractor" who was simply doing his duty protecting the family property.

Later we all sat down and really laughed. But it took a while for us to calm down. We could have blown away Mr. Buffalo, but that would have brought pain down on us from the higher levels. It would also have put the farmer out of business at least temporarily. Don't you love it? I'm still laughing, but I guarantee you that none of us were laughing at the time.

To Vietnam and the Field

Tom Mercer

July 16, 1967. Once we circled around Bien Hoa, Vietnam, getting ready to land, a weird feeling came over me. I took a good look down, and it appeared to be a huge mess to me. Tanks, men with M-16s, and a haze over the city. We were headed to the 90th Replacement Center to get our assignment. This area is where we were assigned to our unit. My friends from Jump School were assigned to the 173rd Airborne Brigade already but for some strange reason, I did not go with them; they never called my name. I went to the Big Red One – 1st Battalion, 18th Infantry, Charlie Company

Di An. Home of the 1st Battalion, 18th Infantry.

in Di An. The men who worked at Long Binh said I was going to a battalion that saw a lot of action and wished me luck and told me to be careful. Little did I know that most of my buddies who went to the 173rd Airborne Brigade were killed at Dak To. I have thought about those guys every day since 1967.

Once I got to Di An, we went out on ambushes and a few patrols for training, with no action at all. We also had to pull LP and perimeter guard. I guess this was getting us ready for the field because we had a lot to learn. Come to find out, all new guys were called "FNGs." This is where we met Ervin Fox, the supply sergeant for Charlie Company. He gave us all of our clothes that we were going to wear out in the field, our web gear, and our rifles. Ervin also told us not to lose or mess up the weapons because we would have to pay for them. I was thinking to myself, "What the hell is he saying?" Then he smiled and said, "Just be careful and watch out for yourself."

Finally it was time to go out to the field. Brandon Doyle and I were shipped to Charlie Company together. I was assigned to Lima Platoon and Brandon to Mike Platoon. Brandon was a good guy and

1967. Left to right: Johnny O'Conner (point man), Ken Gardellis (point man), and John DeSouza, Lima Platoon.

turned out to be a good soldier. I am glad to say he made it out of Vietnam alive and is doing great.

Now my nerves were really working. The chopper ride out to the field was my first, but it was a great ride and what a view. Come to find out, that would be our main transportation for getting around other than walking.

As we flew over the NDP, I got a big lump in my throat. I kept wondering if I could do this and make it back home. But the dust and smell made all that go away, and I focused on what was happening down below. As we landed, I saw 15-20 men running out to the Chinook and start to unload the supplies. A few guys got on the chopper, and I saw them shaking hands with some of the other guys. I asked why and was told they had done their time and were now going home. Even now I remember that day, and I was glad for them going home, but I thought about that for the entire year I was there.

Once the Chinook took off, I was introduced to the guys in Lima Platoon, First Squad. I got instructions, and they took me to my bunker that would be home for a few days. I couldn't believe the way we were going to be living. I could see it was going to be a real tough year. I met the platoon leader and the squad leader, Sergeant Mac (Patrick McLaughlin), who had been in the country for about six months. O.C. (Johnny O'Conner), a point man, was the second person I met, and he was from Florida. Bob Duncan, the machine gunner, was from California. I was assigned to the First Squad in Lima Platoon. They gave me six hand grenades and C-Rations for the day, and I was ready for action.

The next day I went on my first patrol. I felt like everyone in the patrol knew what was going on except me. It was different than what I expected. I believe Robert Norris was the first person I talked to a lot. He had been in the country for about one month so he was an FNG just like me. Robert made it easy for me to get adjusted to the field by telling me things I needed to know, such as I would have to learn how to ration my water so I would always have some left over at the end of the day.

Nothing happened on the patrol that day, and I was glad. After patrol we went to the NDP and rested for a while and then got our night assignment, which was going to be my first LP. That means if the VC were coming toward us, we would be the first ones to hear or see them…maybe!

A few weeks had gone by, and I was getting used to what was going on. I noticed the old timers were starting to talk to me a little,

Song Be. 1967. Tom Mercer receives his M-14, which means he would be walking the point.

asking where I was from and so forth. It seemed like at first they would stay away from us new guys, like we were jinxed. Come to find out you had to earn their respect and show them you knew what to do. I just did my job without complaining; after all, I had volunteered for Vietnam.

After a few weeks of nothing but patrols, I found out I didn't like being in the back of the formation. I thought I would like being up front where I thought the action would be. But I still wasn't sure about walking point since I had not done that yet. I had heard a lot about being a point man: things like it was one of the most dangerous jobs in Vietnam. I had heard when you get into a firefight, you're chances of survival are very slim in the first few seconds. But I liked knowing what was going on, so I decided to volunteer to walk point. I asked Sergeant Mac who said that would come later, just pay attention to O.C. as he was walking point. So that's what I did, and about two months later I finally got my chance to be up front. I'm sure it must have been a real safe area or Sergeant Mac would not

Di An. Lauren Coleman.

have let me try. The point man has a lot of responsibilities. He has to be aware of everything around him. If it didn't look right, you have to stop and check it out. Of course, it didn't hurt to pray every now and then.

Once I became point man, I finally got my M-14, the best rifle in Vietnam at the time. The majority of the point men all carried M-14s, and we were a special breed of men. You had to have a lot of fight in you and know how to put fear aside when it came time to fight. We earned everyone's respect. I can't remember where or how I got my M-14. A lot of the M-14s were taken off choppers, and we left an M-16 with ammo in its place. That's how good the M-14 was.

In the middle of August a new group of guys joined us in the field. Lauren Coleman was among that group; he later became one of my best friends. He was from upstate New York. Coleman and I were in a lot of firefights and battles together in '67-'68. Coleman became an RTO for the platoon leader and was a good solider for Lima Platoon.

Left to right: Kenny Gardellis, Tom Mercer, Point Men of Lima Platoon, 1967-1968

Later in August, Kenny Gardellis joined us in the field. Kenny was from Manhattan, New York. He would become my best friend while in Vietnam and a point man and a real good one. We were in a lot of firefights and battles together and a lot of ambushes. These guys made being in Vietnam a lot easier to deal with.

After about two and a half months, everything started to go as planned, and we hadn't had anyone killed in our platoon yet. We had been on search-and-destroy missions, patrols, and ambushes, but little did we know what was in store for Charlie Company in the weeks to come.

Jungle Rot or Rotten Jungle

Bill Sullivan

One chapter, one name: Lieutenant Colonel Richard E. Cavazos. Lieutenant Colonel Cavazos was the battalion commander for the 1st Battalion 18th Infantry. I had just arrived in country and was finishing up my quick course on the VC, the NVA, the terrain, and Vietnamese customs. It was also a period for trying to adjust to the temperature and humidity. Being from the South may have given me a leg up on the other guys trying to acclimate themselves. On our last day before deploying with our unit, Lieutenant Colonel Cavazos met with us and told us we fight like old ladies. He said he would prep the area with artillery and airstrikes before we go in, but it was up to us to finish the job and reach our objective. He said our loved ones sent us over here alive and in one piece, and it was his job to return us in the same condition if at all possible. It was reassuring knowing that he would support us in any way he could.

His care and concern for his men would become more evident in the months to come. I was in the NDP and was examining my arms and legs. I had sores just smaller than a quarter from my elbows to my hands and from my knees to just above my boots. I had not complained or seen a medic. There was stuff more serious they had to attend to. It would appear that I was shamming (goofing off). Lieutenant Colonel Cavazos came by and noticed my condition. He came over and asked if I had seen a medic and I answered no. He immediately summoned one. He told the colonel that he did not have the proper medicine to treat me. All he had was equipment and medicine to treat gunshot and shrapnel wounds. The colonel told me we would be deploying to Phu Loi in a few days and he wanted me

1968. Bill Sullivan (left); Platoon Sergeant John Kia (right), Mike Platoon.

to get a shower immediately and report to the base medical unit for treatment.

As luck would have it, my squad leader ordered me to pull berm guard while everyone else went for showers. I did not argue as I was still the FNG. I did as I was told and waited for them to return. They took their time and returned late in the afternoon. I was relieved of guard duty, and I headed for the showers. When I arrived, I was informed that the showers were closed for the day, and I could return tomorrow for a shower and clean fatigues. I wasn't the only one turned away. There must have been 40 or 50 men from other units. We all were returning to our units, dirty and disgusted. As I was walking down the road, a jeep pulled up beside me, and there sat Lieutenant Colonel Cavazos. He said, "I thought I told you to get a shower and report to the medical unit." I explained my situation and he said, "Get in." I jumped in, and he had his driver do a U-turn and head back to the showers. Upon arriving he was met by a sergeant who informed him that the lieutenant had closed the showers and taken the key to the lock on the conex that held clean clothes. The colonel ordered the sergeant to open the showers. The sergeant stammered that he would go find the lieutenant and return with the key and permission to open the showers. Lieutenant Colonel Cavazos convinced the sergeant he didn't want to waste his time and that if the lieutenant had a question, he could see him tomorrow. The sergeant immediately opened the showers, and Lieutenant Colonel Cavazos took a pick axe from the jeep and broke the lock on the conex. The other soldiers saw this and began to cheer and came running back to the shower area. It wasn't just me, Lieutenant Colonel Cavazos cared about all the infantrymen and the way they were treated in the rear areas. He knew and understood the life in the field and in combat.

I got my shower and was treated at the medical unit. I was taken off patrol for a few days as I was getting tetracycline shots and having my bandages changed twice a day. I returned to my unit and assumed my duties, but I never forgot Lieutenant Colonel Cavazos's commitment to me. Sometimes it is the small gestures that mean the most, and you never know how your actions will impact someone's life.

Oddball

Bill Sullivan

Every tour is made up with some oddball stuff and here is one story. Three new guys and I had joined the unit. That first night we were preparing for insertion the next day. As we tried to sleep, we kept hearing the artillery firing H and I. We knew they were prepping the area for us. One of the new guys asked the question we were all thinking, "Do you think we are going to die tomorrow?" I responded with a "shut the hell up and go to sleep." We inserted into a small LZ and moved into the jungle. We went on patrol and moved into a large enemy base camp that had been evacuated. There were unexploded cluster bombs in the area. We had to be careful of booby-traps as well as our own ordnance. The cluster bombs were yellow and stood out against the tropical jungle. We marked them with toilet paper anyway. The VC had disarmed some cluster bombs and turned them upside down with the fins sticking up and used them as candle holders. Upon return to our NDP, we prepared for ambush that night. It was rainy, which made hearing footsteps especially hard. Each rain drop falling from one leaf to another sounded like enemy movement. Visibility was about 15 feet, and any combat was going to be up close, personal. I was given the assignment of first guard duty because I was new. That was standard procedure because the new guy was always scared and would stay up most of the night pulling everyone's guard for them. The next morning we retrieved our Claymores and headed back to the NDP. I was the last in line and had rear security. The man in front of me, another newbie, had a cluster bomb in each hand. He had picked them up on the trail. I said, "I think those are cluster bombs" to which he replied, "No, they are candleholders. I saw some just like these in the base camp yesterday." I couldn't convince him otherwise. Maybe he thought I was trying to get his war souvenirs. By the time we got back to our NDP, the OPs had left for the day. As we entered the NDP, we were met by a sergeant who asked the new man what the hell he thought he was doing. The new man explained what he thought they were, and the sergeant told him to place them gently on the ground and move away. The sergeant tore him a new one for being so stupid and bringing live cluster bombs back into our NDP. The sergeant made the new man fill sandbags and place them around the bombs just in case they exploded. After piling them up about five or six sandbags high, the new man placed toilet paper around the sandbags. That night the OPs came back in and the LPs went out. The next morning the sergeant came by to inspect the bombs. The sergeant went crazy. He started yelling, wanting to know who had taken a crap on the bombs. One of the OPs from the day before admitted it was him but explained that he had been out when the bombs were brought in and only returned to the NDP after nightfall. Upon seeing the sandbags and toilet paper he thought that a latrine had been dug and used the sand bags as a seat to do his business. Funny? Yes, but not if the bombs had gone off.

Lesson Learned

Richard Rossi

I n October of 1967 things began to happen that abruptly changed the way I thought of myself as a soldier. There were ambushes, fire-fights, incoming enemy mortar rounds, and human-wave assaults over the next several months. I was fortunate I was not hurt.

Water plant. February 1968. Rick Rossi sitting down, November Platoon.

In life there are lessons to be learned all the time. The experiences I had in Vietnam made me realize that playing soldier in a make-believe world was quite different from the chaos, pain, and death of real war. Nowadays I rarely allow myself to drift into the world of make-believe. The memory of the real experience of war is always right there to remind me of reality.

Joining the Company

Bob Norris

I remember getting off the big bird and thinking this place is hotter than Fort Jackson. They quickly loaded us up in deuce-and-a-halves for our trip to the 90th Replacement Unit. I already had orders for some helicopter unit and was going to be a door gunner, or so I thought. While at the 90th the orders were changed, and I was then assigned to a unit with the 173rd Airborne Brigade. That wasn't going to be good at all. Even though I wanted to go to jump

school, that never happened, and now I was going to be a LEG in an airborne company. On the day of departure from the 90th we were called into formation, I guess to load into buses or trucks to be shipped to our new units. Orders were handed out, and we were pointed in the direction of our rides. As I settled in the back of the truck with my duffle bag between my legs, I noticed that the drivers were armed with weapons but not us. I started to read my orders and realized that I was now assigned to the Big Red One, Charlie Company, 1st Battalion 18th Infantry in a place called Di An. At first the road was hard paved, later it turned to red clay, and as the trucks went speeding along, the red clay turned into red dust.

Finally we arrived at Charlie Company. We spent the next few days completing paperwork for the company clerk and got our issue of jungle fatigues, boots, socks, web gear, helmets – everything to be a jungle warrior but no weapons yet. Then we had to pack all our other clothes and personal items into our duffle bags and turn that over to the supply clerk who would store it away until our return, a year later if we were lucky.

We were told we would have to go to this jungle training school before joining the company in the field. That never happened. The next day we were to fly out in the resupply chopper to the field. We were issued weapons, M-16s, and some ammo. Morning came and we were called to formation. I noticed several others who were also in formation who did not have all the equipment that was loaded on us new guys: just weapons, fatigues, and helmets, and they entered the chopper, a Chinook, first. Now it was my turn as I struggled up the ramp in the rear with a load on my back. I started to stumble as I tripped on some tie-down hooks. I was headed for a fall into the floor opening of the chopper where all I could see below me was red clay. Then all of the sudden I stopped. Someone had grabbed hold of my harness on the web gear and thrown me back against the wall of the chopper. I looked up and staring me in the face was this guy with horn-rim glasses, the biggest toothy grin, the biggest ears, and the bushiest mustache I had ever seen. He said to me that I wasn't getting out of going to the field that easy and to sit down. I had just met the man they called O'Be – Sergeant Robert O'Brien.

We flew into the company NDP, unloaded the chopper, and waited for someone to tell us what to do next. There he was again. "Well, what are you waiting for? Get moving." I just looked at him and asked, "Where am I supposed to go?" After a second he motioned to follow him. We got to an area where several men were sitting around either cleaning their weapons or eating. O'Be went over to another sergeant, said something to him, and then walked off. The other sergeant came over and asked my name and where I was from. He told me to take off all the equipment I was carrying and sit down. I was told as soon as the lieutenant returned, he would get up with me. I watched as O'Brien started hugging and shaking hands with all the other guys sitting around like a big reunion. I found out that O'Brien was the squad leader for Second Squad and that he had just returned from R&R. After a while the lieutenant showed, we talked, and then he assigned me to carry ammo for one of the M-60 machine guns. At first that didn't bother me as I thought, What the heck? Next day we went out on an operation, my first, and sure enough I was carrying ammo. After a couple of clicks in the bush, that one can of ammo started taking its toll on me, but I couldn't put it down; so on I went. Later on we stumbled across a VC base camp with a large rice cache. Some of the old timers said the base camp was larger than expected, and it made them very uncomfortable. We were told to destroy the rice in place. The lieutenant told the squad leaders to send him a few men to help with the rice. Being new, of course, I was one of the volunteers. We then proceeded to pick the 50 or 100 lbs. bags up, slice the bag open, and spread the rice on the ground. This lasted for about an hour, and we didn't put a dent in that cache. After a while someone decided it was time to leave the area, but we still hadn't destroyed all the rice. We moved back to the NDP. That evening we could hear explosions in the distance in the area of the rice. Someone decided to blast the jungle as it was thought to be a major VC base camp.

Surprise Visitor. Sometime late March or early April my squad had just returned from night ambush and was cleaning weapons before going to grab something to eat. David Deeter, Lieutenant Mello's RTO, came over and told me to get to the chopper pad. Oh no, what I have done now, I thought. I grabbed my weapon and steel pot mumbling, "Why do I have to unload choppers?" as I walked toward the center of the NDP.

As I approached, I didn't see any choppers. I did notice a couple of guys talking near the pad. One was dressed in standard fatigues, had a .45 on his side, and was wearing his helmet. His side was to me, and as I got closer I thought he looked familiar. The other guy said something and pointed in my direction. When he turned toward me, I almost dropped my jaw. I did recognize him; it was my father – "Oh my God. Dad, what are you doing here?" I asked. I ran up to him, grabbed his hand and squeezed as hard as I could. He wasn't into hugging.

My father was stationed in Thailand at the large air base north of Bangkok. I didn't know what he did there. He was a warrant officer. He told me that on a few occasions he would fly into Nam. He decided to try to locate me. I don't know how he did it, but there he was. We sat down on some sandbags and talked for what seemed like an hour. But a chopper was on approach, and he informed me that he was leaving on it. This time he hugged me and told me to keep my ass down. As I watched the chopper lifting off, kicking up red dirt, I waved to him; he waved back. I sat back down and cried, then got up and walked back to my squad. Mello looked at me and grinned.

Rockets' Red Glare. In January 1968, Charlie Company had just returned to Lai Khe after an operation for what we thought would be a few days of rest. After getting showered and into clean fatigues, the lieutenant called for all the squad leaders. We were to pull perimeter security, but there was a catch. We were to secure the local village that was within Lai Khe, and we were to detain any REMF who tried to leave the village and turn them over to the MPs.

My squad moved to our assigned area near some old French buildings with a couple of sandbagged bunkers nearby. We set up near the bunkers with our backs to a large open field and on the far side of the field, about a click away, was the tree line. I split the squad into two-man teams, including me and my radio man. I told them that at least one was to be awake at all times. We weren't in the jungle; so it didn't seem that critical. A couple of the guys used the bunker as a place to bunk down. We were well into the early morning hours and still no activity from the village. I moved around to the men to be sure someone was awake. As I approached one of the bunkers, a loud screeching noise like a freight train was over my head. When I looked up, all I could make out was a long, stove-like tube floating overhead. I even thought I could make out the markings, it was so slow. They were Russian 122mm rockets. Several more were screaming into Lai Khe and landing somewhere beyond us and the village. My guess was the airstrip. My guys were all into the bunkers, and at the same time I noticed movement from the village. I had this ugly gut feeling that Charlie was in the village and was now coming out to attack Lai Khe, but no, they were the REMFs flying out as fast as they could. So much for our mission to collect them and turn them over to the MPs. Of course we let them go. I raced back to the radio and was told to get ready for a ground assault I passed the word to my squad, checking with them to be sure they were ready. By now illumination was in the air and counter fire was going out. Then everything got quiet. We waited; nothing happened. As soon as the sun came up, we joined the company. Everybody was gearing up, and then we headed to the airstrip.

There we loaded the choppers and flew off. We landed somewhere outside of Saigon. That's when we found out that the VC and NVA had launched the TET Offensive throughout the country.

To the Field

Kenny Gardellis

I leave Di An and make my way to meet my comrades-in-arms who are guarding the village of Quan Loi. We load up and get on choppers, and flying above it all I can see mostly rice paddies and small villages, groves of fruit trees, lots of palm trees, but the further north we go the redder the earth gets…red mud in the monsoons and powdered fine red dust in the dry season that gets blown in your face by the choppers. The terrain up north is more isolated with more jungles, rubber tree plantations, and smaller villages. We land inside the barbed wire surrounding the base of Quan Loi and find it to have some creature comforts like showers and permanent bunkers that are deep, well-made, and ringed around the whole base camp. Not too shabby compared to the days and months ahead where we are flown out near the Cambodian border and have to dig our foxholes, fill sandbags, and cut down trees to use the trunks as logs for overhead cover. The permanent bunkers here near Quan Loi are in under the shade of a grove of trees, which I think are rubber trees, but I'm not sure. I am assigned to Lima Platoon.

As a new guy I start out as a rifleman. Eventually I am switched to doing different jobs, including ammo bearer as well. The ammo can is heavy, and I put a rifle sling on the handle so I can carry the box, which hangs down about waist high. Going up and down hills or wading through creeks and trying to get up the slippery bank on the

Song Be. 1967. Right before Loc Ninh, NDP.

other side while the heavy can pulls me backward. The can swings and knocks you off your stride while walking, sinks you deeper in the mud while crossing rivers and streams, and hurts if you swing back and hit your hip or leg. Aside from being an ammo bearer, I eventually carry the PRC-25 radio for the lieutenant for a while and hold some other positions., However, the Big Red One values its point men. If you are good, a little seasoned, and acclimated to the food and weather, they will put you up front. Eventually I am appointed to this important position, and it defines much of my experience in Vietnam. During my first months in Lima platoon, however, our point man is Johnny, a tall, blond guy from New York via Key West. He has long legs, intelligence, a sense of humor, no fear, and a mind both cautious and quick. He has a special sixth sense that allows him to sense when danger (i.e., VC) is near. He can see signs of enemy activity on the trails and in the woods and jungles too. Johnny is color blind. That gives him a different perspective on seeing the enemy.

My first meeting with the guys of Lima Platoon goes well – no jokes or wisecracks – but I notice that since we all came into country at different times, there is a core group in Lima that has been here six months or more. A lot of them have been in firefights and seen some crazy stuff prior to my arrival in August. The battalion has just finished a major offensive operation up near the Cambodian border called Junction City. It's documented in books as a major offensive, important to the US troops because they went out to destroy the COSVN, the VC main command post believed to be located underground in Tay Ninh province near a mountain called Nui Ba Den, the Black Virgin Mountain. So there is still talk from the seasoned vets like: "Remember Junction City?" "Remember Shenandoah?" At this the guys will all stop what they are doing and look into the eyes of survivors like Johnny, Sergeant Mac, O'Be, Duncan, and Porky. It was a trial by fire. When soldiers survive enemy bullets and booby traps, they count their blessings and then start joking about something mundane, like cookies from home, and they all laugh – that laugh of relief – that moment of "Holy crap, we cheated death back then."

Joe Boland's Story

Joe Boland

My name is Joseph Boland. I spent a year in Vietnam from June 1967 until June 1968. I was assigned to Charlie Company, 1st Battalion, 18th Infantry, 1st Infantry Division. I saw a fair amount of combat in that year. I saw young men wounded and killed in battle. I was able to avoid being wounded while there and came away with an impression that it was just a matter of inches that might spare your life or limb. And danger wasn't just limited to combat. This series of stories, or better yet, "experiences," as my fellow Charlie Company comrade Bob Quimby says, pertains to the close calls I experienced.

Joe Boland.

Tail Rotor Blades. I have wanted to thank a man whose name was Henry Bibby ever since I left Vietnam. Henry was a young black man from somewhere in the South. He was muscular and scrappy. Henry worked in supply with me. We were unloading a Bell UH-1 Iroquois helicopter (Huey) that was resupplying our field location. Upon landing, Hueys kicked up a tremendous amount of dust and dirt from the prop wash of their rotor blades. I was trying to hold the supplies and food in Mermite cans down. I had my eyes closed because the dust was so bad. I was somewhat bent over to hold lids on and such. Suddenly I was tackled and hit the ground hard. Henry was on top of me. He had been watching what was happening and saw the tail rotor coming toward me and knew I didn't see it. He spear tackled me and landed atop me as the Huey's tail whipped around and over the top of both of us. I wouldn't be here if not for Henry. He only had an instant to react, and he risked his own life to save mine.

Ordinarily, the pilots took off in a straight line forward. One of the soldiers on the supply detail thought he saw a door gunner get into the pilots seat. He may have been receiving instructions from the pilots on how to fly the Huey. I was told they did that in case the pilot and copilot were wounded and could not fly themselves out of a dangerous situation. Whatever the case, THANK YOU, Henry. I am indebted to you for saving my life. May God bless you wherever you are today.

Claymore Mine. Charlie Company seemed to rotate through different field night defensive positions and base camps. Our home base was Di An, 8-10 miles northeast of Saigon. Other base camps were Lai Khe, Quan Loi, Phuc Vinh, and Song Be. We also had NDPs off Highway 13. Charlie Company manned one of these Highway 13 NDPs. From these positions we would sweep the highway with mine detectors each morning and secure the road throughout the day. At night we would move back into the NDP and man the perimeter through the night.

The NDPs were usually circular and fortified with bunkers that provided overhead cover and firing ports that overlapped with adjacent bunkers to ensure that fields of fire were entirely covered. Sleeping positions directly behind the bunkers were sandbagged. Ponchos were used as tents over these sleeping areas to keep the rain out. These particular NDPs were built for a single company to man. Charlie Company always dug in. Without bunkers we were too vulnerable.

The perimeter is where we put the crappers and piss tubes. They were out beyond the bunkers. A crapper was fashioned out of ammo crates with a hole over a ½ barrel. Privacy wasn't important. We'd pull the barrels out from under the ammo crate toilet and would have to put kerosene or gasoline into it to burn the crap. It was always crawling with maggots. Sometimes lindane powder would be there to dust on the fashioned toilet seat as a measure to kill lice and crabs.

Bringing in supplies outside of Di An, Joe Boland's job.

One night I had been constipated and was sitting on this makeshift toilet just before dusk. Men were getting ready for the night. I could see men beginning to string their Claymore mines out. Claymore mines were a convex-shaped antipersonnel mine with hundreds of ball bearings imbedded in a plastic explosive called C-4 and covered in plastic. They would blast outward from the detonation point. I didn't want to be out there if anyone got careless, but I was taking longer to get a needed bowel movement and had been in enough discomfort to persist. I watched as a soldier positioned his Claymore on its legs to get the correct elevation. I was off to the side at maybe a 45-degree angle from him and not directly in the Claymore's path. He rechecked its position and put the blasting cap in it and straightened the wire back to his position. He tested the detonator a number of times without first attaching the wire. He then attached the wire to the detonator. When he was finished, he tossed the detonator down onto his bunker's sandbags. I watched as

31

if in slow motion. When the detonator hit the sandbag, it sent an electrical charge through the wire and to the blasting cap. The Claymore mine detonated. It scared the s⊛t right out of me, literally. I stood up and looked all over my body for wounds. Adrenaline had kicked in, and I was on overload. I don't know how it could have missed me except for the angle the Claymore was set in relationship to me. An alternative view was that it wasn't my time. My memory has faded as to what happened afterwards.

M-79 Grenade Launcher (thump gun)-High Explosive M-79 Round. During the latter part of the week of the Tet Offensive, which began on January 31, 1968, I had returned to Vietnam from R & R leave for a week or so in Hawaii, where I met my wife. The timing was just right. I left Vietnam before the enemy's offensive began. When I left, everything was as normal as could be in a war zone. The civilian airline flight to Hawaii was routine and no cause for alarm. Upon returning I saw a completely different story. I could see from the air while landing that a lot of destruction had occurred. There were burnt-out armored vehicles littering the sides of the runways. Buildings were damaged. It was a mess.

I was unaware of what was going on in Vietnam the week I was away. I had not watched television or picked up a newspaper. I was oblivious to what had happened during Tet. It was difficult to leave my wife and return to Vietnam at the end of my R & R. I had five months left in my one-year tour. Now I saw that the war went on without me and from the look of it, very badly, too.

I rejoined my company. They were securing a power plant or water treatment plant in Saigon. This NDP was like an R & R. We had hot showers and flushing toilets. One of my company comrades was killed while on operations in this area. Every time they left the perimeter, they had some hostile activity. Having just returned to Vietnam, my alertness level was still stateside and not in sync with being back. The company radio telephone operators, Bob Quimby,

1967. Fire base for Charlie Company Oscar Platoon, NDP.

Tony Kulakowski, John Watson, a medic, and forward observer's RTO named Poncho decided to play a practical joke on me. Headquarters' platoon was set up under a canopy of the main building that we were securing. They knew that when our mortars were fired, the sound would be amplified as it reverberated through the open canopy. They were monitoring the radio and heard when Oscar Platoon was ready to fire. I wasn't paying attention to the radio and was unaware they were about to fire. When they fired, all of them ducked for cover because the noise reverberated under the canopy and sounded like incoming and not outgoing mortar fire. Poncho was the instigator.

When it was actually incoming, you could hear the distinct "bloop" sound of their mortar rounds leaving the tube. Sometimes that "bloop bloop" could wake you out of a sound sleep, and you would scramble for cover before they hit. I got so I never took my boots off at night. I would just loosen the laces. I learned that lesson after tripping on some wire in my bare feet during one of my first mortar attacks.

One day I had been ferrying grunts in a ¾-ton truck. I pulled into the compound at the power plant with them in the back. I hopped out of the cab and was near the back of the truck. One of the grunts was carrying an M-79 grenade launcher. The breach and the locking mechanism were open on the thump gun, and when he jumped and hit the ground, the high explosive round dislodged from the breach and fell to the ground. It hit on the primer side of the round and shot straight up into the air and came back down within a few feet of all of us. First Sergeant Ward looked around at everyone and then asked me to pick it up and dispose of it. My comment was, "You've got to be s°°°°ing me, Top!" He certainly didn't pick it up. After some debate, someone said the explosive part of the shell has to go through the rifling in the barrel of the thump gun to be armed. I didn't know that. From my standpoint it was live, and but for the grace of God, we were safe. Someone took care of that round that day, but it wasn't me.

Top must have liked me because at another place and time, he had me police up armloads of C-4, blasting caps, and hundreds of feet of detonation cord that the engineers had wrapped around trees at an NDP to clear fields of fire for the company's bunker positions. On this particular day they had detonated quite a number of explosions and caused some consternation at headquarters. They had left this ordinance unexploded, and it was rigged in series to blow with detonation cord strung between C-4 charges for one instantaneous explosion. I followed the detonation cord to the farthest point, removed the blasting caps, and worked my way back to the perimeter with C-4 and detonation cord wrapped all around me. Top was good enough to send a couple of my comrades out to help me tote the explosives back.

Was It the Enemy or Not? While still operating out of the Thu Doc power plant, the company conducted daytime patrols and night ambushes in the area. Captain Phil McClure was the company commander. He was a no-nonsense military man. Sometimes he had a short fuse, but I would learn later that he had told his platoon leaders about his temper and told them that after his flare up, he would not hold a grudge. I didn't know this. I was asked to drive from the Thu Doc power plant to Di An to pick up some supplies, including batteries for the starlight scope. I remember driving back to Di An, but I do not recall who, if anyone, rode shotgun. This was frequent, hostile enemy activity during Tet. While en route to Di An, there were dead enemy bodies alongside the roads. Other dead could be seen in the distance. I remember driving on a stretch of road where buildings butted right up alongside the road. Suddenly, about 15 to 20 Vietnamese, heavily armed and dressed in black pajamas, stepped from behind a building right onto the roadway just in front of me. They were carrying their weapons but not in the ready position. My heart jumped right up into my throat, and I thought my ticket would be punched that day. Because I was driving, my M-16 wasn't in hand and useless anyway from behind the steering wheel. All I could do was wave and smile, and I didn't have to think to do that; it just happened. Surprisingly, they waved back. They were across the road in a blink and between buildings on the other side just as I was passing. I thought I saw RPGs on their shoulders and AK-47 rifles. My mind has replayed this over and over again throughout the years. I am uncertain now what parts of it are imagination or embellished to the point of becoming my reality. There may have been friendly's in the area, but this was Tet, and I wasn't about to stop and ask them. They may not have wanted to draw attention to themselves and their position. I'll never know if they were friendly or not.

I got to Di An for the supplies including the Starlight batteries and made my way back to the power plant somewhat unnerved. While in Di An I must have gotten caught up in something that now escapes me, and I arrived at the power plant after the company had set out for a night ambush. Two Vietnamese children were wounded the morning after the ambush before daybreak. I carried some guilt for a lot of years because of this. They may not have been wounded if I had made it back with the Starlight batteries before the ambush departed.

I have since had the occasion to meet then-Captain Phil McClure at a reunion in April-May 2010 of our company comrades who were with us at the time. He spoke with empathy and compassion. He talked about the incident where the two boys were wounded. He said that he had called a medevac chopper in to evacuate the boys for emergency medical care. He said that he had gotten word that both survived. After he spoke, the burden of guilt I carried for so many years began to lift.

Malaria. I remember being in a rubber tree plantation. We had what's called a stand down, where we came into a base camp and had the occasion to unwind, drink some beer, and be away from the harrowing aspects of war for just a day or two. The day after, I had a horrible hangover. It seemed to persist beyond the first day and into the second. When I drank water, I'd just vomit it right back up. I became dehydrated and had a fever. On the third day, the hangover was still there, or so I thought. I went to see someone at a field hospital.

They took a blood sample from my arm with a syringe and took my temperature. Another GI and I were sitting on a curb, both sick as dogs. Medical personnel came over to us and discussed our cases as if we weren't even there. As they discussed our test results, one said to the others, "One of these men has spinal meningitis and the other has malaria." He didn't say which one of us had what disease. I sat there and wished or prayed that I didn't have spinal meningitis. A trainee in my basic training unit had contracted spinal meningitis, and I knew enough to not want that outcome.

As it turned out, I had malaria. It was diagnosed as Falciparum malaria, which is the worst kind to contract and has the highest rates of complications and mortality. When it was discovered that

I had malaria, I recall being placed in an ice bath to get the high fever down. The field hospital was in a tent with wooden pallets for a floor. I was given intravenous fluids with quinine. I recall that I was still vomiting off the side of a bed through the slats of the pallets onto the ground before they brought me a bucket. I couldn't keep either food or fluids in my stomach for a week or more. I had no desire for food. I was so sick I didn't care if I would make it. After some time had passed, I remember looking at an orange on my breakfast tray and not being sick from just looking at it. I ate it and kept it down. That was the turning point for me. I think I was sent to the larger evacuation hospital near Di An and then on to Cam Rahn Bay to recuperate.

Cam Rahn Bay was like an in-country R & R, but I didn't get to take advantage of anything there. I was too weak and my appetite still hadn't resumed as normal. I had now been away from Charlie Company for a couple of weeks and felt some insecurity about losing my job.

I saw a lot of injured soldiers while in the various hospitals and felt guilt over contracting malaria. When I was able to get up and move about, I saw even more wounded soldiers. We had few walking wounded if any in the field, and the number of men in these central hospitals troubled me.

After recuperation in Cam Rahn Bay, I was shipped back to Tan Son Nhut Airport. We were brought down to the tarmac to await a flight. When we arrived, it was too late to catch a ride to Di An. Normally buses ran there up to a certain time of day. Two other GIs were trying to get to Di An besides me. Each of us had been in the hospital. We didn't have any money or a place to stay. There wasn't a coordinator waiting for us when we arrived. It seemed to me that we waited all night. I was still very weak, and it was cold out there.

We encountered a civilian taxi driver who agreed to take us. Since we didn't have money, he wanted us to buy him some things at a post exchange, which required a ration card. None of us were armed, except that one of the GIs carried a Bowie knife. We bought the cabbie a large electric fan, cigarettes, and whatnot. He kept trying to dicker for more. When we went to his cab to discuss, the GI with the Bowie knife had it drawn and at the neck of the driver. We were headed for Di An shortly after. The driver spoke limited English, and he became very hesitant en route. He said he could not go any farther as we approached a South Korean compound outside of Di An. A Korean soldier stopped us, and the taxi was turned around. We exited the vehicle as the Korean pointed up the road. We started walking, none of us armed. It was near dusk. I don't know if the Koreans' radioed to Di An or whether someone saw us walking up the road, but a jeep arrived from Di An to pick us up. A young lieutenant chewed us up one side and then the other. He drove us to the gate opposite of where my company's compound was. It had been an ordeal to get back.

When I reached my barracks, I was completely exhausted. I fell into a bunk with a mattress and blankets and was fast asleep. I was startled awake by a GI who had a knife to my throat screaming, "What the hell are you doing in my bunk?" I was able to get up and find another bunk. It had no mattress or blankets, and I laid down on the springs and shivered the rest of the night. It was good to be home.

Rocket Day. Others in Charlie Company have written about the day when we were sent on a patrol to find the location from which the enemy fired 122-mm rockets into the base camp of Lai Khe. These rockets left a sizable crater on impact. On this particular day, the enemy saw us and began to fire the rockets. The grass surrounding the rockets was tinder dry and a sizable blaze erupted. Other rockets were lying in the fire. The ground here was extremely hard. I was on the perimeter of the fire that reached waist high. I had a D-handled shovel and was attempting to throw what dirt I could scrape up onto the fire. If the rockets exploded in place, the kill zone would have been devastating. In the heat of the fire, I became dehydrated quickly. The fire was going to cook the rockets off in a matter of time and anyone as close as me would not survive. Suddenly a couple of our guys ran into the fire and picked up the rockets. They didn't think about them being booby trapped. They may have had to make multiple trips to get them to a safe area out of the fire. My brain didn't work like their's. I didn't even consider running into the fire. Yet it was the only way to survive that day. Thomas Cone and Tom Mercer were the ones in the fire that day. Thanks, guys, for saving my life.

Jump from Huey into LZ. There was one particular LZ that we were ferried into in lifts of five Huey helicopters at a time. This particular LZ was the same one where engineers had rigged explosives to trees to clear fields of fire. Before our assault it had been prepped with artillery. As a result, felled trees, stumps, and debris were scattered all over the entire area. Because of the debris, we were told that the choppers would not touch down and that we would have to jump.

I was in the second or third lift of five Hueys and in the second chopper of that lift. As we approached the LZ, I was standing on one of the skids and holding onto the doorframe. We came in low, and the first chopper touched down in the spot where it was supposed to be. The first chopper may have blocked our pilot's view, so when we reached the spot where we were to touch down, our pilot saw a downed Huey from a previous lift in our spot. Consequently he began to lift up just as I committed to jump. It happened in just seconds. I was pulled up well above and beyond the downed Huey. Because of the debris on the ground beneath me, I had to stick a landing like a gymnast does. With the full weight of my combat gear, I had a bone-jarring impact. I managed to get out of the gear and drag it over to where HQ was setting up. I was in the best shape of

Lai Khe. 1967. Preparing to chopper out to some unknown LZ.

my life; my core body muscles were strong. I had severe back aches and the medic gave me Darvon tablets, which I didn't take because I didn't want to become drowsy.

Shortly after I was discharged from active duty, I began to have recurrent back problems that necessitated many different therapies: chiropractor adjustments, traction, spinal blocks, and cervical vertebrae fusion. Every disc in my lumbar/sacral spine is either bulging or herniated, and two vertebrae have been fused. The nerve root was severely compressed, and I have lost strength and flexibility in my upper extremities to the point that I cannot change a light bulb overhead. I take medication for nerve pain, muscle relaxants, narcotic pain medication, and anti-inflammatory medication. I have radiculopathy in my arms and hands. I often get spasms, numbness and tingling in my hands and fingers. I am convinced that I sustained a compression type injury to multiple levels of my cervical, lumbar, and sacral spine from this incident.

I cannot believe how much this hindered my quality of life. I haven't played golf in over 30 years. I quit riding my Harley and sold it because it became too difficult to operate the hand controls. The constant vibration from the handlebars made my hands numb. My wife has done the lifting for years, and it is becoming more troublesome as she ages.

Point Man

Tom Mercer

Being a point man was a hell of a job. We should have received hazardous duty pay, because it was a job most of the guys did not want to do. When a firefight starts, the point man's life expectancy is only a few seconds from the first shots fired. That was why most of the guys didn't want to walk point.

Point men were a special breed of men who were a little crazy but smart. They had to have fast reaction times when the s°°t hit the fan. Sometimes you didn't have a lot of time to determine who's the bad guy and who's the good guy. That's why there were free fire zones and no fire zones. No fire zones meant there were civilians around, so we had to be careful who we were shooting at. Free fire zones meant there should be no friendlies at all around, and you had better shoot when you see something. I know when I was on point and came up to some friendlies, they had better identify themselves fast. If they had a weapon, it didn't matter then; they were shot or shot at.

Most of the point men in C Company carried M-14s and not the M-16. The M-16 jammed way too often and being on point you didn't need that. The M-14 was a lot heaver but, to me, it was a great weapon. I loaded my magazine with duplex rounds that had two leads in each round – one went high and one went low. The magazine in my weapon was filled completely with tracer rounds, this way

1968. Left to right: Lon (Smitty) Smith, Kenny Gardellis, Tom Mercer, and Lauren Coleman relaxing after a patrol.

the damn VC could see them coming, which would make them get down and give me a few seconds to get to a safe spot – or cover. The other magazines had tracer rounds every other round, so we could see where our rounds were going, and could adjust our fire. These rounds anybody could use, not only point men.

A lot of point men in Vietnam used shotguns, which I tried, but I went back to the M-14. I was always worried about the shells getting wet and not firing, and the M-14 would tear up some jungle if you were firing into it. Some point men even carried the M-79 grenade launcher with canister rounds, which was like a big shotgun round. But I was worried about coming up on some VC at close range and the M-79 would not be as effective as the M-14. The M-79 round had to travel so far before it would explode, and point men didn't need that.

A point man had to be alert to everything around him at all times. He's not only looking for VC, but booby traps, bunkers, snipers, and anything that didn't look right. If you got into a big hurry, you were asking for trouble. Getting back to the NDP was always good, but it's better to get back alive than in a body bag. We never rushed.

We had some great point men in Charlie Company in the year I was there, guys like Kenny Gardellis, Bob Norris, Buffalo Boy, Bill Sullivan, O.C., Patrick McLaughlin, Lon Smith (Smitty), Tommy Strano, and Thomas Cone. I know there are more, but the names aren't in my head now. One guy could not walk point every day; it would drive him crazy. Most of the time we rotated platoons on point, and the platoon leader would pick the guy to take the point.

Bill Sullivan who walked point for Mike Platoon was a great point man. If he saw someone having to walk point, and they had a wedding band on, he would take their place. He said he couldn't stand the thought of having a married man with a family and kids killed while on point.

Most good point men always complained about having to walk point, but once we were up front we were all business. I guess it was just being in Vietnam. We thought we had the right to bitch. Point men had to earn the respect from all the men in their platoon and squad, but once the men realized you knew what you were doing, then they trusted you, and it made them feel a lot better on patrol. Every point man had his own way of walking point, and he hoped it was the right way and worked out well each time.

The Price of a Smile

Patrick McLaughlin

29 October 1967. We were in an inverted "V" formation moving through the rubber trees. First Squad, Lima Platoon, Charlie Company had the point. The "old man," Dogface Six, wanted his best on point. Military intelligence

Loc Ninh. Nov 1, 1967. Patrick McLaughlin preparing to take the point for Dogface.

(MI) confirmed an enemy unit was positioned some 500 meters or so north of our night defensive perimeter (NDP) and Dogface Charlie was going out to pick a fight.

I cut my teeth on point and still walked it, though not as frequently as before they made me a squad leader. There were two of us on point because of the formation. Johnny O'Connor was to my right about 10 meters, and Lima fanned out behind. Mike Platoon was behind and to the right, November Platoon behind and to the left. We headed toward the jungle line in the distance at rubber's end. This is where they would spring the ambush and my mind sought to come to grips with this really bad situation. Assuming, that is, that MI was right this time.

"Gooks!" Johnny sounded the alarm, and he and others hit the ground, save me. I'm focused on the jungle line and see nothing. "Johnny, where?"

"Mac, get down! They're right in front of us."

My eyes scrolled down the row of rubber trees from the jungle line less than 100 meters distant, and there he was. In an irrigation ditch running across our path, less than 15 meters from me in my row, I see the barrel of a ChiCom machine gun pointed right at the bulls-eye on my chest. It resembled a howitzer. Behind that barrel, in total command of the gun and the situation, was an NVA soldier. He waited till my eyes met his, pleased by the shock he saw in my face. He lifted his head, broke out in a big sinister smile and pulled the trigger. His smile was gone. He and I both knew that his target was a dead man.

October 24, 1967. Having had the good fortune of taking the first R & R allotment to Sydney provided to the Dogface Battalion, I arrived back in Di An (pronounced "Zian"), where Charlie Company, 1st Battalion, 18th Infantry, 1st Infantry Division, the "Big Red One," had our permanent hooches.[1] I will always be a goodwill ambassador for the Aussies.

The 1st Battalion, 18th Infantry was constituted in the Regular Army in 1861 and organized at Camp Thomas, Ohio. The 18th fought in many campaigns during the Civil War, including Chickamauga, Chattanooga, and Atlanta. The 18th fought in the Indian Wars, the War with Spain, the Philippine Insurrection, World War I, and World War II before service in Vietnam.[2] I learned this history upon arrival at Di An.

On June 8, 1917, the 18th was assigned to the 1st Expeditionary Division, later re-designated the 1st Division, and still later, the 1st Infantry Division (the "Big Red One"). During World War I, the 18th fought at places with names like Cantigny, Aisne-Marne, St. Mihiel, and Meuse-Argonne.

---

1 R&R was an informal term for leave for purposes of "rest and recuperation." See the List of Terms for explanation of other terms.
2 The 1st Infantry Division was on occupation duty in Germany from 1945 to 1955 and did not participate in the Korean War, 1950-1953.

Song Be. Right before the battles at Loc Ninh. (Ray Etherton, reclining; John May, looking back, Lima 5, KIA on Nov 2, 1967; Paul Zima, Lima 6, WIA and dusted off on Nov 2, 1967; and Bob O'Brien.)

In World War II the regiment fought in North Africa (Algeria, French-Morocco, and Tunisia), Sicily, Normandy (starting at Omaha Beach), Northern France, Rhineland, Ardennes-Alsace, and Central Europe. At the war's end, the 18th served as the Honor Guard for the war crimes trials at Nurnberg. The Regiment's motto is In Omnia Paratus ("In All Things Prepared"). There is no infantry division, Army or Marine, with a record finer than that of the 1st Infantry Division. The 18th Infantry is certainly a primary reason why that is so.

As soon as I hit the company area, I told the NCOIC to get me on the next chopper back to the unit. He did and I joined my squad – 1st Squad, Lima Platoon, on October 24th. We were at Song Be. I have been the squad leader since early April, a rare private first class (pay grade E-3) squad leader. By July 22nd, the one year anniversary of taking my oath in the U. S. Army, I had been promoted to sergeant

Fall 1967. Getting ready to saddle-up and move out.

(E-5). I was 20-years-old. Fourteen grunts made-up the squad when I became "Lima One." It was my job to lead them and get them home in one piece.

The Squad Leader. The job of infantry squad leader in Vietnam was a bitch; it was a tough assignment, and one of the most important. Squad leaders ran the point elements. Squad leaders took out the 12-man ambush patrols at night, isolated from the main force, in the bad guy's territory.

The squad leader served as a buffer, and a sounding board, between the brass and his men, striving always to strike the right balance between being a buddy to his squad members and the first-line superior who led them into harm's way. He covered for his guys when they deserved it and kicked their ass when they needed it. If the squad leader lived long enough, he was the most experienced combat leader in the infantry platoon. The squad leader was in the field for the duration of his one-year tour while officers generally rotated in and out at six month intervals – it was said to "punch their tickets."

In the Dogface Battalion we ran two squads to a rifle platoon, so an M-60 machine gun team of four (gunner, assistant gunner, and two ammo bearers) was part of my squad. Lima's "1st Gun" was Bob

Duncan, from California. If you were in a fight, you'd want "Dunc" (aka "Hollywood") on your side. The M-60 machine gun was an ass-kicking weapon that fired the 7.62 "NATO" round. Squads usually numbered a dozen guys, more or less.

Second Squad was led by Robert O'Brien ("O'Be"), from Massachusetts. Call sign "Lima Two." O'Be, who was older than the rest of us grunts, was on his second tour in Nam. On the first he was a clerk in Saigon. He kept pestering his superiors to let him transfer to the infantry. They did, and he ended up with Dogface Charlie (although we were "Duchess Charlie" in early 1967). O'Be, who was blind as a bat without his glasses and couldn't see all that well with them, joined Charlie Company in the field several weeks before I joined in early January 1967. He was prior service, having spent four years on active duty with the Air Force, and six months active with the Marines.

It was very rare for anyone to volunteer to transfer from a rear echelon position to the infantry. Not many wanted to be in the infantry, for obvious reasons. All other jobs were seen as a step up and out from being the basic light weapons infantryman – a "grunt" – occupational specialty 11 Bravo. When other slots would open up, such as door gunner on a helicopter, grunts would volunteer. Anything to get out of the field, particularly if one was getting "short;" that is, closer to the end of a one-year tour of duty. O'Be was on his second tour in '67. He volunteered for a third straight tour in '68 and became a staff sergeant (E-6) advisor to a Vietnamese Civilian Irregular Defense Group (CIDG) in Tay Ninh province. He sought a fourth consecutive tour, but the Army told him to go home. It was O'Brien that laid the handle "Mac" on me, which stuck throughout my tour and beyond.

One of O'Be's two team leaders was Ray Etherton ("Porky"), from southern Illinois, a top-notch combat soldier. Porky didn't look the part, but he damn sure acted it. In early summer 1967, during a bad fight, Dogface Six called in an air strike. A couple Air Force fighters dropped some stuff close to our perimeter (where the enemy was) near where Porky and I stood side-by-side. A sizable chunk of shrapnel struck Porky in the back of his right arm cutting out a large "v-shaped" chunk to the bone. A couple inches over and his arm would be severed. The pilot learned that we had a "friendly fire" casualty and flew over low and slow, tipping his wings in salute. Seen as a decent gesture, the grunts wave back, although Porky may have called him a name. Porky was dusted off and spent a couple months in a hospital in Japan. He rejoined Lima Platoon before the fights at Loc Ninh.

Long Live the M-14. Upon checking in at Song Be, I collected my M-14 rifle from the grunt who I favored to leave it with and gave him back his M-16, which I had carried to the rear with me. In our platoon, you never took your M-14 out of the field. Our M-14s were coveted. They were revered. We humped seven M-14s in Lima Platoon. You could dump the M-14 in a mud puddle or a rice paddy, pick it up, wipe off the goop and fire away.

Steadfast and reliable, the M-14 was a man's best friend in Lima Platoon. The M-16s had proven unreliable. They would jam, requiring us to connect the cleaning rod pieces and use it to unseat the jammed cartridge from the chamber – much like Davey Crockett had done loading his long rifle. Eventually, in 1967, the problem was fixed, but good men died or were maimed for life because of this deficiency. The designers should have gone to jail. Better yet, they should have been sent to Nam to carry their M-16s – with us.

Many Enemies in Nam. The enemy wasn't the only problem we faced. The mosquitoes were relentless and big as dragonflies. We bathed in "bug juice" – Army-issued insect repellant. We reeked of it, but those damn bugs still came around. And with them came malaria. If the malaria was severe enough, the grunt was choppered back for a hospital stay, then back to the boonies.

We feared the red ants and always took pains to insure we did not bed down on those devils. They were mean and so aggressive that they would leap out of trees or bushes to get to you.

Perhaps even more hated were the leeches. We were magnets drawing the leeches to us and often couldn't challenge the bloodsuckers for fear of exposing our position. So we gritted our teeth and tried to think about anything other than which body part or orifice was the next target. When we returned to the perimeter in the morning, we stripped down and our buddies helped in finding and removing the blood-sated bastards we couldn't easily reach. The art was to pull out the leech without leaving its head in whatever body part it was attached. Two methods worked best – pour bug juice on the critter or, my favorite, touch it with the lit end of a cigarette.

Then there were the snakes, many kinds, some big, some small. The big ones, the pythons and boa constrictors, really weren't too much of a concern. You could easily see them. But the little ones were a different story. We knew one as the "bamboo viper," more notoriously referred to as the "two-step snake" – once bitten, you take two steps then you die. The tarantulas were big and hairy, and liked to climb in the poncho liners with us to stay warm. I saw three types of scorpions – black, red, and white. We also had tigers, panthers, and wild boars cross our paths at various times.

Some enemies were of our own making. "Friendly fire" was always a hazard. It came from "short" or off-target artillery and mortar rounds, airstrikes, rounds fired from our own guys during firefights, ricochets, and sometimes bizarre mishaps and human error.

We made up names for some infirmities, like "bamboo poisoning." Risks to point men in particular were the cuts sustained on hands and arms as you chopped your way through the bamboo and foliage. At one point, I developed an infection on both hands that caused them to swell-up to twice normal size. It was so painful that I couldn't put my hands in my pockets, and it was a problem just to hold my M-14 in a firing position. I needed help to open C-ration cans – I couldn't use the "p-38" can opener. I went to Lima's medic, John Houchins, who wanted to send me back to the rear for treatment. I didn't want to leave my squad, so we consulted a doctor who gave me a shot of penicillin – worked like magic.

Not all illnesses or deaths occurred on the spot. Some enemies lay dormant for years, only to emerge in various deadly forms. The infantry spent a lot of time in the jungle – the very jungle that some geniuses decided to defoliate in order to deny our enemies its concealment. They used "Agent Orange."[3] We grunts lived in it, patrolled in it, bathed in it, and slept in it. Many have died from it since. Children of Vietnam veterans have been born with deformities causally related to their fathers' service in Nam. More vets will continue to die from its effects. After decades of wearing blinders, our own Veterans Administration decided that a number of illnesses are linked to exposure to "Agent Orange": diabetes, type 2; ischemic heart disease; Parkinson's disease; prostate cancer; Hodgkin's disease; B-cell leukemia; peripheral neuropathies; and other illnesses and disorders with, no doubt, more to be identified. Given the option, we would have taken our chances straight up against the men we faced.

October 26, 1967. Lima platoon had a relatively new "Lima 5," the platoon sergeant, a lifer from Illinois, SSG (E-6) John May. He integrated seamlessly into the ebb and flow of Lima, which had a uniform mix of FNGs,[4] more seasoned soldiers, and "short timers"

---

3 Agent Orange is the name given to a blend of herbicides the U.S. military sprayed from 1962 to 1971 during Operation Ranch Hand in the Vietnam War to remove foliage that provided enemy cover. The name "Agent Orange" came from the orange identifying stripe used on the 55-gallon drums in which it was stored. http://www.publichealth.va.gov/exposures/agentorange/basics.asp.
4 New replacements in Vietnam were often called "FNGs" for "f*****g new guys."

who would be leaving soon. May earned our respect quickly, which wasn't always easy. Grunts who have fought and bled together are discerning and unforgiving critics of anyone, of any rank, with whom they must go into harm's way.

We also had a new platoon leader, Lima Six, Lieutenant Paul Zima, who, like SSG May, was from the Chicago area. Zima called the squad leaders up to his CP bunker to tell us that Charlie Company would leave Song Be in the next day or so to go back to Di An for two weeks.[5]

When O'Be and I briefed our squads, spirits were lifted. Di An was "safer" than the areas where we operated, so it was a "treat" to head back there. Before leaving Song Be, one of Lima's newer guys, Tom Mercer of Florida, received his birthday present early – he got an M-14.

October 28, 1967. Early in the morning, we broke camp and headed over to the airstrip. We choppered out of Song Be on to the quite sizable 1st Brigade base camp at Lai Khe, where we were placed in a rubber grove well inside the perimeter. To our surprise we were given the next couple days off – no patrols, ambushes, listening posts (LPs) or any combat duties. No one could recall that Charlie Company had ever been in such a stand-down setting. Fine with us.

October 29, 1967. About 0400, my radio/telephone operator (RTO), PFC Lauren Coleman from New York, touched my shoulder and stepped back. I'm instantly awake and reach for my M-14. "Mac, not a problem. Are you awake?"

"Yes, what's going on?"

"Charlie Six wants to see all platoon leaders, platoon sergeants,

---

5 In 1967, the Army created a fourth rifle company, D ("Delta") Company, for all infantry battalions in Vietnam. This allowed three companies to be in the field, with one in reserve, securing the base camp. It was our turn to rotate back as Alpha and Bravo Companies had done earlier.

Song Be. Oct 28, 1967. Preparing to chopper out to Lai Khe. Engaged in battle the next day at Loc Ninh. McLaughlin, Lima 1; Coleman (in background), Lima 1 Kilo; Etherton, Team Leader, 2nd Squad Lima; Rucker, Rifleman, 2nd Squad Lima; Cone, Rifleman, 1st Squad Lima; and O'Brien, Lima 2.

and squad leaders right away. Something's up."

We made our way in the dark aided by a few flashlights that we could use since we weren't on the perimeter. Charlie Six had everyone's attention. Bill Annan, from Baltimore, was a West Pointer and a serious company commander. He had been with us a couple months, straight from the spit-and-polish of a line outfit in Germany.

"Gentlemen, we will be moving out at daybreak to the airstrip, where we will travel by plane to Quan Loi and from there by chopper to Loc Ninh. Loc Ninh is northwest of here along the Cambodian border. A Special Forces camp and a CIDG compound at the airstrip were attacked early this morning, and the situation is uncertain. The fighting is ongoing."

We learned that Dogface would be inserted somewhere between Loc Ninh and Cambodia to cut off reinforcements coming in and to take away escape routes for enemy units then engaged. We were to set-up in a rubber plantation that had been designated as a "no fire zone."[6]

The planes were waiting and we landed at Quan Loi in no time. Someone was anxious for Dogface to get in the action so we didn't do the customary "hurry up and wait." At Quan Loi we loaded into Chinooks. This was unusual, as we hit hot LZs in Hueys. This telegraphed that we weren't going in "hot."[7]

We carried everything a grunt needed to hit the boonies and dig in "Big Red One" style – which nobody did like us. This included axes and chainsaws to cut overhead cover, and beaucoup sandbags. This, all in addition to the accouterments of war: fragmentation ("frag") grenades, smoke grenades, white phosphorus grenades, claymore mines, four canteens of water (you did not want to run out of water), extra ammo (for me, 20 magazines), a few carried LAWs, C-rations, poncho and liner, protective mask, some carried air mattresses (extra weight), entrenching tools (shovels), first aid items, and whatever personal things one desired to carry (camera). This did not include underwear or socks. Only new guys fussed about such items that were wholly unsuited to the climate. Experienced grunts soon learned that they were extra weight, particularly when waterlogged.

---

6 Dogface had experienced "no fire zones" before, particularly around Quan Loi with its rubber plantations. The French overseers still lived in large homes around Quan Loi. One had a swimming pool and clubhouse which, unfortunately for them, the officers had commandeered. The rumor among us grunts was that if we damaged any rubber trees during operations, the U.S. would pay reparations to the French owners. They, in turn, would pay taxes to the enemy in order to be left alone. Rightly or wrongly, we envisioned the French as a mere pass-through of US dollars to the enemy. In the simple- mindedness of the guys who actually fought the war, this was FUBAR.

7 A "Chinook" is a CH-47 double rotor, medium lift, cargo helicopter capable of carrying about 40 fully loaded troops. Because Chinooks were much larger and much less maneuverable than UH-1 "Huey" assault helicopters, they were rarely used in infantry air assaults into known enemy locations.

We were loaded down – those humping M-14s carried more weight since the 7.62 ammo was heavier than the M-16's 5.56 ammo.

We were quiet now. The complaining and grab-assing was over. None of us knew what was in store for Dogface Charlie. Each man kept his own counsel. We were a unit of primarily 11 Bravos heading, once again, into the unknown. As we chopper overhead, who below was watching? Were we seen as the enemy – liberators – invaders – allies – or just strange men from a far off place: Mars, perhaps?

I looked around at the guys, really kids mostly. Some were old enough to vote. Enlistees and draftees, different colors, religions, and economic circumstances now bound together by the arbitrariness of this war. We were sent here by "we the people" to fight, to survive or not – who knows? Time will tell.

One day at a time and today is October 29th. I've got two more months of hard time, then back to the world. One of my team leaders looks over – our eyes met, he smiled, I nodded. A solid soldier, John Willett, was from New Hampshire. Quiet, reliable, ever quick to lighten the mood, John is a man you can count on.

On this day in 1967, none of us could ever have imagined that if we returned home we would be met frequently with disinterest, disrespect, vilification, and blamed by some for the increasingly unpopular war by the very people who sent us off to fight it.

Loc Ninh. The Dogface Battalion air-assaulted into a clearing alongside a large rubber plantation – the Plantation des Terres Rouge – about two-and-a-half miles west of the airstrip at Loc Ninh and about an equal distance from the Cambodian border. The task of digging in was at hand. All of Alpha and Delta Companies were in the rubber, as was Charlie, except for Lima Platoon. Our first bunker was positioned outside the rubber starting with 2nd Squad. Lima's CP was in the center and 1st Squad positions extended to the battalion's reconnaissance platoon (Recon), which ran the half-circle back to the rubber.

My bunker linked with Recon, and we were recessed in the rear position. Our "first guns" (M-60 machine gun) position was to my right and front. The bunkers, typically manned by threes, were staggered one in front and one behind then one in front forming, ultimately, a "wagon train" effect. Lima and Recon sheltered the 105mm howitzers that were likewise dropped in by Chinooks in the open space between the rubber tree line and the semicircle formed by the joining of the two platoons. The 105s provided fire support to the Special Forces camp and the Loc Ninh airstrip, and the 105s positioned at that location provided fire support to Dogface. Charlie Oscar's 81mm mortars were also in the clearing positioned to rain bad news on the enemy.

At my position were my RTO, Coleman, with the PRC 25 radio. Each squad leader, platoon leader, and company commander had a radio and, thus, an RTO. First Squad's senior point man, Johnny O'Connor ("O.C.") from New York, was the third man at the position. An infantry platoon is full of characters, and Johnny was certainly one of those. Blond, and but for the New York accent, he could have been singing with the Beach Boys. Smart and mouthy, O.C. was one of those soldiers who expected things to make sense. In combat, making sense was a luxury we rarely enjoyed.

The Point. Point men required a keen eye and a natural instinct. They can't teach that, either you have it or you don't. Throw in courage and luck, and a point man had a chance to survive.

The terrain was often jungle, thick brush, and heavy bamboo that required the infantry to move in a single, extended file "snaking" through the jungle terrain. The first man in that file, leading the way – out front and exposed like no other anywhere – was the point man. He was the first to confront the enemy, the first to get cut down. If hit, he was either alone, or among just a handful of other men in a position to engage the enemy.

One of the all-time most miserable days of my life was a day on point moving from one NDP to establish another. We were humping everything we carried to the field – 80-plus pounds of weapons, gear, and stuff. It was one of those occasions when I decided to take the point, and none of the guys complained about it. It was hot, unbelievably hot. It was 120 degrees Fahrenheit with humidity, it seemed, to match. The medics were handing out salt tablets like they were life savers.

The bush was thick, full of bamboo and heavy vines that I hacked through with a machete. It was a wall of impenetrable green. The going was so physically demanding that it took every ounce of my self-respect not to just quit. The presence of an entire infantry battalion in single file behind me pushed me onward. It was so difficult I removed my ruck-sack and passed it back for someone else to hump. At one point, as dumb as this sounds, I physically couldn't work the machete and carry my M-14 so I handed my weapon to the man behind me and asked him to stay close. "Don't leave me naked standing here without my M-14 if we walk into some shit." A couple of the guys gave me sporadic breathers.

The platoon leader, Lima Six, asked me to hold up. I took a knee to catch my breath and pulled out a canteen. After several moments Lima Six told me to cut a new azimuth to take us to a clearing not far off. He said something about a dust-off coming in. This made perfect sense to me. I might climb on that chopper too, if I could make it that far.

We made the clearing and secured it for the Huey inbound. We popped green smoke. Looking around, I didn't see the man or men who no doubt suffered heat stroke or exhaustion and required emergency evacuation. We soon learned it wasn't one of us. Well, it was one of us, but not of the two-legged variety. It was the German Shepherd scout dog traveling with us. The dog was unable to dissipate the oppressive heat and was dying. The Army got it right and sent a Huey to save it. The dog's handler carried the dog out to the chopper as soon as it touched down, laid our canine comrade gently within, climbed on and off they flew. No doubt, though exhausted, the dog looked down on us grunts gazing longingly at the Huey and felt sorry we were left behind. It was a miserable day, one not fit for a dog.

1130 hours, October 29, 1967. We were digging in "DePuy" style when the Lima Six RTO, David Estus, from upstate New York, radioed over, "Lima One, Charlie Six wants to see you and Lima Six at the command bunker." I acknowledged, threw on my shirt, grabbed my steel pot and M-14 and headed into the rubber. What a difference the shade made. The sun was brutal out in the open where we were digging in – this seemed like air conditioning. "Digging-in" was a project, and we did our share of griping about the time and physical effort required to do it right. No infantry unit dug in like the BRO in late 1966 and 1967.

The division commander when I arrived in country was Major General William E. DePuy. He directed that all infantry units in the field construct bunkers with overhead cover and firing ports, with bunkers staggered front, rear, front, rear and so on. A foxhole must be deep enough for two men to stand upright so that they could fire their weapons from the standing position. The third man sat in the well and overlooked the bunker.

We filled sandbags with the dirt from the foxhole and placed them around the foxhole, leaving space for the two firing ports at the front corners and at the center rear, the well, which is the entrance with steps down into the bunker. Typically, three sandbags stacked one upon the other would suffice. Utilizing the axes and chain saws we humped, or that were flown in after we hit the perimeter (yes, including five-gallon cans of gas), we cut logs for overhead cover. We made a lot of noise but the choppers dropping us off made even more – so we couldn't keep our presence a secret. Two layers of sandbags were placed upon the logs and extra dirt went in front to form a berm. At each firing port sandbags were placed going out at a 45 degree angle to suppress the muzzle flashes so that the only way for the enemy to see the weapon firing was for him to be in the line of fire or "kill zone." Camouflage was added to aid in blending in with the environment.

The "V" positioned sandbags of the bunkers in the rear positions were carefully lined up so that the men firing from those portals could cover the bunkers to the front. Their lines of fire took in the forward bunkers themselves and the space in front, between, and beyond them. This way, if the enemy overran the forward bunkers, the men in the rear were positioned to fire in defense of their buddies. Again, the only way for the enemy to see where the firing originated was if they were in the "kill zone."

Some smart guy figured that we grunts should also have some extra protection, so a directive was issued to extend rows of sandbags to the rear of the bunkers forming a rectangular sleeping area. This permitted two men to sleep in some security from shrapnel and small arms fire while the third man stood guard. I trust that General DePuy, and his successor General Hay (who continued to improve the defensive fighting positions), got the credit they deserved because their BRO defensive fighting positions saved many lives and inflicted severe damage whenever the enemy sought to overrun them.

1200 Hours, October 29, 1967. Arriving at Charlie Six's position, Captain Annan laid out a map and began his briefing. "Sergeant Mac, you are here with the platoon leaders because the old man wants your squad on point." The "old man" was Dogface Six, Lieutenant Colonel Richard E. Cavazos, commander of the 1st Battalion, 18th Infantry.

Cavazos was an ROTC Texas Tech graduate who played football in college. As an infantry platoon leader in Korea, he won the Distinguished Service Cross, the nation's second highest combat decoration. Already he had made a name for himself in the Army as an up-and-coming infantry officer. For those of us in the Dogface Battalion, the "old man" was special. His command instincts were exceptional, and he loved his grunts, who would and did follow him anywhere. The 1st Battalion, 18th Infantry became the Dogface Battalion when Lieutenant Colonel Cavazos took command in March 1967 and changed the call sign from "Duchess" to "Dogface." We were then, and will forever be, Dogface soldiers.

Military intelligence reported a known enemy unit 500 meters or so north of our current position, advised Annan. We were going out to find the enemy. Charlie Company would go out in an inverted "V" with Lima's 1st Squad on point, Mike behind to the right, November behind to the left. Oscar Platoon manned the mortars and was standing by inside the perimeter. Charlie Six doled out various instructions, map positions were coordinated, and the known presence of the enemy nearby was hammered home. There were no questions.

Charlie Company saddled-up and when Lieutenant Zima gave me the word we headed north deeper into the rubber. I was humping 11 magazines in addition to the one loaded in my M-14, duplex rounds, every fourth or fifth round a tracer, three fragmentation grenades, and two smoke grenades. Typically, the magazines were loaded with 18, sometimes 19, rounds. The duplex rounds were two in one. Two projectiles (rounds or bullets) made up each 7.62 cartridge, with the second bullet fitted into the hollowed-out back end of the front projectile. When the trigger was pulled, two projectiles headed down range, separating as they traveled. The M-14 could bring smoke, big time.

Because we were in an inverted "V" instead of the usual column formation, Lima had two point men: O'Connor on the right and me on the left. The closest man to me was Coleman about five meters to my rear. The "pace man," Lonnie "Smitty" Smith from Nebraska, was right behind him. O'Connor, who also carried an M-14, was 10 meters to my right. First squad was behind and to my right with Mike Platoon fanned out behind. Second squad was to my left with November Platoon to their rear. Lima's Five and Six, and their RTO, were positioned near the center of 1st and 2nd Squads.

The day was full sun, no clouds, and oppressively hot. The rubber trees provided shade with intermittent sunlight cramming through. Line of sight ahead, along the rows of rubber, was unimpeded, and also left and right down those rows as one walked along. Irrigation ditches, deep and wide enough to conceal an enemy force, ran with the contour of the terrain.

After 300 meters or so, the sounds of Companies A and D back at the NDP fell away. They were digging-in while Charlie Company patrolled, so when we got back, we would have to hustle to get our bunkers up and on line.

It was clear, easy going, no need for a machete here. O.C. and I made frequent eye contact to stay connected and in sync. Our pace was measured and controlled. We both expected something but were quite unsure what it would be. Ken Gardellis, from Pennsylvania, brought up the rear of 1st Squad. Between Gardellis and the point were Willett, Mercer, Smith, Beal, Cone, Brandon, Martin, Duncan, Myrick, Biser, and Coleman, all alert.

1230 Hours, October 29, 1967. Out now 400 meters from the NDP, I could discern the "jungle line" at the end of the rubber maybe 200 meters ahead. If I were the enemy, this is where I would spring the ambush. There was cover and the ability to move around undetected to better position themselves to hit us as we approached in plain view. How experienced – how good – are these guys? What would I do in their place?

As I stepped over another irrigation ditch, looking left then right, it occurred to me that the enemy could spring an ambush from there, but the jungle line remained my first choice. The ditches would provide concealment to the enemy but not give the undetected maneuverability the jungle wall provided. Johnny and I were talking back and forth about the possibilities and the enemy's options. I cautioned 1st Squad to be alert. We were 500 meters out and less than 100 meters short of the jungle line.

"Gooks!" Johnny shouted loud and clear. I scanned the jungle line for movement and saw nothing. I looked to my right. O.C. was in the prone position, as were others. I shouted, "Johnny, where?" O'Connor indicated movement in the ditch to his front.

"Mac, get down! They're right in front of us."

Quickly shifting my line-of-sight down the row from the jungle line, I spotted him. Less than 15 meters directly to my front in the ditch that ran perpendicular to my path was an enemy soldier. I'm standing upright; there is no cover between us. The enemy soldier has a clear line-of-fire. To my left was a rubber tree 10 feet away, to my right the open space between the rows of trees. The soldier was positioned behind a machine gun on bi-pods that rested in an indentation he had dug straight out from the ditch. This concealed the gun somewhat with the barrel hovering right above ground level. His left hand was cupped on the top of the weapon, right finger on the trigger and his head cocked right looking down his barrel pointed right at my chest. He was patient, waiting until I saw him.

The Smile. When our eyes locked, he lifted his head above his weapon and smiled at me. He wanted me to know that I had just gotten myself killed, that he would be the last person I saw on this earth. His smile was big and sinister. The smile conveyed that it was important to him that his target, a stupid, clueless American knew that his life was over. He was pleased that he had waited to kill this enemy soldier rather than cutting him down without warning. The look of shock on this American's face said it all – he did know that he was a dead man! His patience had paid off. He would tell this story to his sons and their sons.

The command to pull the trigger began in his brain and the

Brandon, Martin, McLaughlin, and Smith of Lima Platoon at Loc Ninh after battle of November 2, 1967.

impulse moved down his central nervous system out to the periph-eral nervous system, traveling through the protection of the myelin sheath to his right trigger finger. The impulse arrived instanta-neously. As the soldier squeezed the trigger, his smile disappeared. The first shots fired at the Dogface Battalion at Loc Ninh cut through the stifling heat of this early Vietnam afternoon.

His smile shocked me as I realized my fate. The thought that I would die today, in this place so far from home, exploded in my mind. My immediate emotion was anger at myself: "You idiot, you're wasted. He's got you."

My life then passed before my eyes. Warm memories of grow-ing up with my mother, father, and sisters, and of places we had lived all brought contentment to my mind.

Then, a deep sadness enveloped me. I knew the impact my death would have on Mom. She would not be able to move on. Even today, these memories trigger emotions that imprinted them-selves on me in that blink of time.

Just as the inevitability of my certain death settled in, another part of my brain analyzed the situation. Instinct assumed command. I may die here today – but not willingly. Subconsciously, from deep in my brain somewhere, instinct overrode the logic of diving for the cover of the nearest rubber tree. Instead, I dove right, to the open space away from the only available shelter available. As I dove right, the NVA soldier fired a burst of six to eight rounds into the space to my left. He'd counted on his target diving for the only cover available. He miscalculated. He missed me.

I fired a round before hitting the ground and, by sheer luck, the bullet impacted just in front of my adversary. He jerked his head back and I was on my knees, firing again. He ducked down in the ditch and I now had the advantage, firing a few rounds into where his head had been. I'm thinking grenades. Pulling a frag grenade from the canteen pouch I used for extra grenades, my dilemma was whether I lay my weapon down in order to pull the pin and throw. If the enemy soldier fired a burst at me while I'm on my knees fumbling with a grenade, it would be…very unpleasant.

Inwardly, I chuckled at the image of Hollywood "heroes" pulling grenade pins with their teeth. Were it only that simple! Putting my M-14 down and hugging the ground, I pulled the pin with my finger, armed the grenade, counted "one thousand one, one thousand two" and threw. I overshot the ditch. I had not adjusted for the adrenalin factor. I hollered for Coleman to toss me some grenades. He tossed me his two, which gave me four frags. My second grenade was on target, right in the ditch and timed perfectly. It blew as soon as it landed.

We have movement in the ditch to my right in front of O'Connor and down the ditch from him. Are they in the ditch to the left of me? "O'Be!"

"Yeah, Mac"

"I've got this guy under control but don't know if they are in the ditch to my left. Can you and Porky check it out?" They both responded "Will do." They moved swiftly, flanking to the left and while they were doing so I lobbed another grenade. This one landed right on the far edge of the ditch over the location of the gun and exploded. Not as effective as a direct hit, it was nonetheless lethal.

O'Be and Porky hollered back that the ditch was clear up to the position to my front. I rushed the ditch and jumped in. To my surprise, there were four NVA soldiers at the machine gun position and not just the man with the sinister smile.

Looking immediately to my right down the ditch my vision was blocked by a large ant or termite mound right in the ditch about 25 feet from me. It appeared that my grenades had killed the four NVA soldiers. As I bent down to double check, an AK-47 on automatic opened up on me from the mound. The cracking blasts of the rounds seemed to suck the air out of my lungs. I don't know how they missed. Pivoting right, I fired several rounds into the mound, letting this character know I was alive and kicking.

While pivoting, I stepped on a body and was shocked to see he was a Dogface. John Willett had jumped in the ditch to my left, away from the mound. The AK-47 rounds that passed me hit John. At this point, we were the only Dogface soldiers in the ditch. Willett hadn't wanted me to be alone in that ditch. O'Be and Porky were down to the left some distance; the remainder of Lima was sitting tight.

John had a bad head wound – a geyser of blood was squirting up 18 inches above his head from a deep gash over his ear – he was bleeding to death. Stepping over John to reposition myself to fire again at the mound, I stuck my left thumb in the hole in John's head and yelled, "Medic up." The bleeding slowed significantly.

I didn't wait for the NVA soldier behind the mound to pop up and unload on us; I placed a couple rounds into the mound to remind him that this may not be his best move. "Medic," I yelled out again. Doc Simpson, the company medic, who accompanied us on this patrol, jumped into the ditch and took over with John Willett. From South Carolina, Doc Simpson was one of those fearless medics a Dogface could count on to be there whenever the call came.

Climbing out of the ditch, I fired a couple rounds past the mound and above the ditch that I could see from this vantage point. Quickly inserting a fresh magazine I shouted, "Johnny, see if you can get some grenades into the ditch. I've got a good angle on him if he pops up." O'Connor, not adjusting for the "adrenalin factor" either, lobbed a grenade over the ditch. His second toss was right on the money. It blew an NVA soldier up out of the ditch and deposited him on the edge. A loud cheer rang out from the guys near enough to see the action. O.C. and others rushed the ditch firing. Three NVA soldiers were in this position, now all dead. One of the three had fired the rounds that struck down John Willett.

Back in the ditch, I lifted the machine gun that nearly carved me up and tossed it out. Someone standing over me fired down the ditch. Looking up, I saw Porky, his M-16 in rapid fire. He spotted enemy soldiers down the ditch. "I nailed one, then a second as they were running down the ditch," Porky yelled. Small arms fire, theirs and ours, grew more intense as Mike Platoon battled the NVA in the ditches off to our right. O'Brien, setting down his M-14, picked-up "my" machine gun and opened up on the NVA down the ditch line. Standing and firing from the hip in bursts of 3 and 4, O'Be exclaimed, "Mac, this is beautiful. It doesn't kick, doesn't ride up; it's smooth."

"O'Be, you dummy, get your ass down." He just looked at me, laughed and kept firing at the NVA he could see off in the rubber. Let them get a taste of their own weapon. Classic O'Brien – one of a kind – an unforgettable character!

As Doc worked on Willett, I directed O'Connor to guide others in getting our wounded comrade back to the NDP. Lieutenant Zima called for a dust-off for Willett, which promptly carried him out of the battle. Charlie Company lost Private First Class Charles Gentry from Cumberland, Maryland, who was struck down in the early part of the battle. He gave his life in service to the nation at the age of 19.

Now, walking into an ambush is bad news, but to walk into the end position of an "L-shaped" ambush is good news – so long as you survive the experience. The enemy had laid out for us the long stem of the "L" winding throughout the ditches to the right of Lima's location. Mike Platoon was pinned down mainly in the open with the NVA concealed in and firing from the ditches. We didn't know it at the time, but we were fighting soldiers from the 165th NVA Regiment.

The NVA launched a counterattack from the east against Mike Platoon's right flank, catching a Mike platoon cloverleaf patrol in a bad position. Paul Tidwell, a Mike grunt who carried an M-14, remembers that he went out about a hundred meters to support his three friends, joining them in a ditch. The NVA assaulted the ditch but were beaten back until an enemy soldier flanked them and fired down the ditch, wounding all four and knocking out two of their weapons. The four wounded men then shared Paul's M-14 and a .45-caliber pistol as they fought off more NVA assaults.

As the crescendo of machine guns, rifles, and grenades quickened in pace and volume, Charlie Six pulled Lima Platoon back toward the NDP and behind Mike Platoon. We then moved east, formed on line and were instructed to move north through the rubber in the on-line formation to flush the enemy out of the ditches and flank the NVA's counterattack on Mike Platoon.

Moving out, I hollered to 1st Squad to stay together so no one was out front and exposed more than necessary. Lima's discipline was on the money – all together, each man doing his part. We moved a couple hundred meters and began engaging and flushing the enemy out of their havens. Mike Platoon greeted the heavy fire from the M-

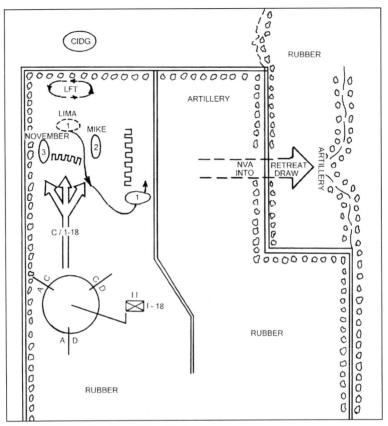

Oct 29, 1967. Battle of Srok Silamlite I.

60, the M-14s, M-16s and "thump guns," the M79 grenade launcher, like manna from heaven. The wounded Mike Platoon soldiers in the ditch were spared the fate that the 165th had in mind for them.

When the NVA soldiers left the security of the ditches, they ran to the east looking for safe rubber. One NVA soldier ran across my row some 50 to 60 meters distant. Looking to his right he spotted me and came to an abrupt halt. He back stepped to the center of the row and glared at me. Slowly, and almost ceremoniously, he raised his AK-47. I stepped left a couple paces and leaned my shoulder into a rubber tree, taking a standing firing position.

The soldier opened up on automatic. Green tracers, seemingly in slow motion, floated to me and by me. For the third time in this fight, I felt the cold, clammy fingers of death reaching for me. My heart pounding, breathing hard, I thought – deep breath, hold, aim, fire on semi-automatic – one shot at a time – keep your composure – squeeze the trigger, don't jerk. Fire! I missed.

The NVA soldier continued to fire on automatic bursts – we both understood that in seconds one of us would die. Repeat the cycle: don't let the situation control, I'm in control. Fire a second round – the soldier stopped, hesitated, dropped to his knees, his AK-47 slipping to the ground. He stares at me. I fired again and he fell to ground.

Another soldier entered my row, sees his fallen comrade, spots me, and runs to the next rows, trying to place rubber trees between us. No time for the firing steps taught in basic training – a quick shot, then a second, leading the racing soldier, this one finding its mark. The 7.62mm projectiles knocked the soldier to the ground. He rolled and bounced up limping. I fired two or three more shots, but only hit trees. The NVA soldier, running and limping, must have praised his deity. For the moment, he had escaped the fate of many of his comrades.

Map Symbols

⊠	Infantry Unit
[•]	Artillery Unit
I	Company, Battery or Troop
II	Battalion or Squadron
X	Brigade
⊣⊦►	50 Cal. Machinegun
O─►	Mortar
◄	Ground Assault Axis of Advance

Delta Company came in on Charlie Company's right, but the fight was over as far as the 165th NVA Regiment was concerned. The NVA were fleeing east, then north, seeking refuge. Dogface Six called in artillery and airstrikes to chase them. I don't know how many were killed or wounded, but they left 24 bodies behind. Dogface casualties were one killed and nine wounded.

The plantation owners no doubt rejoiced at hearing of a large battle in the "Terres Rouge" and hoped to send a large invoice to Uncle Sam for the damage caused by thoughtless GIs. On a positive note, the Washington civilian kicking down a martini at the Willard's Round Robin bar and the rear echelon eggheads tossing down some cold "33s" on the rooftop bar of Saigon's Rex Hotel could sleep soundly that night as 1st Squad, Lima Platoon, honored the rules of engagement and didn't fire until fired upon!

1600 Hours, October 29, 1967. Clean up took a while. We dragged enemy weapons, ammo, and grenades back with us and left the dead for their comrades to collect. We took care of ours and they took care of theirs. I learned years later that this engagement was dubbed the "Battle of Srok Silamlite I."

Lima was exhausted, but we had bunkers to build. O'Brien and I pushed the guys to get it done. Grunts from Recon Platoon and Alpha and Delta Companies came by to hear what went down. The mood was elevated. We won a solid victory, although it was tempered by the casualties we took. A few of the men from Mike Platoon came by to offer special thanks to Lima. One crusty NCO said, "We owe you big time, and Mike Platoon will not forget what Lima did today."

That night, although exhausted, I slept little. The battle kept replaying in my mind. What had I done right, where had I gone wrong? How did I miscalculate where the enemy would spring the ambush? Had the NVA machine gunner cut me down before I saw him, how would the fight have changed? How many Lima men would have died had that gunner and his three comrades opened up on the point element before O.C. eye-balled movement in the ditch to his front? Did the gunner's three mates endorse his decision to not spring the ambush until I spotted him? How many family and loved ones grieved in North Vietnam: mothers, wives, children because of a smile? How many across the Pacific were spared that grief?

Was the smile worth it? It has been to me. It's value to me and others that day – priceless! Its cost to the NVA soldiers in that ditch and others that day – everything! That was the price of a smile.

October 30, 1967. Charlie Company had weapons to clean and resupplies of ammo and grenades to secure. We hoarded extra of both as no one believed that this fight was over. Alpha Company moved out on patrol south and southwest of the NDP. Shortly after noon, all hell broke loose with small arms, ours and theirs, firing hard and fast. Delta Company was committed to the fight and Dogface Six called in artillery and tactical air support. The fight lasted several hours.

Back at the NDP, Charlie Company guys spread out, filling in the positions around the perimeter in case the enemy attempted to assault the Dogface position during the battle.

The 165th NVA Regiment had come back, this time in battalion strength, to avenge the beating one of its companies took the day before. As on the 29th, the NVA used the irrigation ditches – the fighting was intense and at close quarters. First Division after-action reports say 83 enemy soldiers were killed. Dogface casualties were four killed in action and five wounded. This battle is referred to as "Srok Silamlite II." The brass throughout Division and above now all clearly understood that the enemy had stuck around Loc Ninh in order to pick a fight with the Big Red One.

October 31, 1967. Early in the morning, the Loc Ninh airstrip and the Special Forces/CIDG camp were mortared and then assaulted by the 1st and 2nd Battalions of the 272nd VC Regiment. The attacks continued to suggest that the enemy wanted to capture the provincial capital of Loc Ninh prior to the upcoming national elections. None of us suspected that they were part of the build up for the Tet Offensive to be launched in a couple months.

The 105mm artillery unit behind Lima fired, it seemed for hours, in support of the CIDG compound at the airstrip. We didn't sleep much that night. The attack continued well into the morning and the 272nd Regiment withdrew after suffering significant casualties.

The Dogface Battalion caught a breather on the 31st as we manned the perimeter, maintained our observation posts (OP) and listening posts (LP), and conducted ambush patrols. There were no company-sized patrols.

On this day, according to my DD-214, Dogface Six promoted me to staff sergeant (E-6). Lieutenant Colonel Cavazos pinned my new stripes on in Quan Loi just after we left Loc Ninh. To make staff sergeant in 15 months and 9 days was extremely rare. The battalion sergeant major commented that I had the least time-in-service of any staff sergeant in the entire division.

November 1, 1967. It was Charlie Company's turn for patrol and we were joined by one of the other companies – I believe Alpha. The battalion commander would be with us as we all expected more action. I was called up to Lima's CP, and Captain Annan told me that Dogface Six had again directed that 1st Squad, Lima Platoon take the point. This was very unusual for one squad to be singled out when there were so many others that had not yet had their "turn" on point – particularly since we were in a very bad neighborhood. I viewed it as the ultimate compliment but felt we were being asked to carry a bigger load than any other squad in the Dogface Battalion. I wasn't sure how the guys would react. They let me know.

"What, are we the only squad capable of taking the point?"

"Sergeant Mac, this is unfair."

"The 'old man' is asking more of 1st Squad, Lima than of any other squad."

I couldn't challenge any of the comments. When they were done, I simply said that we were given the point because we were the best, and it was vital that the best take the point. I asked one or two of my most experienced guys if they wanted the point, but none did. No problem – I took it. It is best when the man on point wants to be there.

When we stepped off, with the scent of death floating in the air, the two Dogface Companies were in a single-file column and not in an inverted "V." Nor did we have guys out on the flanks, which would be typical. We were expecting trouble and the commander likely did not want cloverleafs out after what happened to Mike Platoon three days earlier.

As I led the column through the rubber, I was in a zone. We had gone a thousand meters or more when I spotted movement to my right and up a hill, which we were navigating around. Holding the patrol column, I froze to take stock. Seconds passed. I zeroed in on the spot where I saw movement. Then I saw them as they stepped into the open at the crest of the hill – three NVA soldiers in uniforms and helmets, wearing backpacks, carrying AK-47s, stood glaring down at me. Fine, I'll take a shot. Just as quickly as I brought the M-14 to my shoulder, they ducked out of sight. "Coleman, I need the radio."

"Lima One to Lima Six, over."

"Lima Six, over."

"I've got three enemy soldiers in uniform and backpacks, with AKs, up the hill to right front, around 2 o'clock. They wanted me to see them before disappearing, over."

"Lima Six, roger, checking with Charlie Six."

"Charlie Six to Lima One, over."

"Lima One, over."

"Lima One, tell me what you have, over." I did as requested, then asked, "Do you want us to go up and check it out, over?"

"Wait one, checking with Dogface Six, out". I didn't wait long.

"Charlie Six to Lima One, over."

"Lima One, go."

"The 'old man' wants clarification. Are you certain that they wanted you to see them, over?"

"Affirmative, they made sure that I saw them before they

ducked out of sight, over."

"Charlie Six to Lima One, over."

"Lima One, go."

"Dogface Six says this is a trap. We are cutting a new azimuth and calling in artillery. Charlie Six out."

We headed off in a new direction while Dogface Six called in artillery on the hill and surrounding area. The two companies of infantry "snaked" its way back to the perimeter. Another day, another patrol – otherwise uneventful – or was it?

It was Lima's turn for ambush patrol, not an entirely happy prospect considering there were a bunch of angry NVA and VC fellows in the area with unpleasant designs on us. It was 2nd Squad's rotation. Sergeant O'Brien and Lieutenant Zima returned from the CP after briefing and O'Be began to brief his guys. I walked over to listen in, as did the platoon sergeant, John May.

It was important to know where 2nd Squad would be positioned for the ambush. Also, 1st Squad had to fan out and occupy the entire platoon front. We could be spread too thin. This coordination was vital because Lima platoon on the north side, and the battalion's Recon platoon on the south side, protected the artillery and mortar positions critical to the overall operation. Staff Sergeant May and I worked out how 1st Squad would "fill-in" the vacant 2nd Squad bunkers.

Second Squad was uneasy about this ambush and rightly so. O'Be calmed his guys, and Porky provided seasoned leadership to the newer soldiers. In the end, you bitched, shook it off, and carried on with the mission. That's what 2nd Squad did. As they lined-up, single file, grunts double checking weapons, ammo, frags, Claymores, I poked O'Be. "Hey, O'Be."

"Yea, Mac."

"Don't start anything you can't finish out there."

"We're just going out to make a little fire, toast marshmallows, and tell some ghost stories. Wanna come?"

"Appreciate the invite but someone's got to hold this perimeter while y'all are out camping. See you in the morning."

As light was fading, O'Be and his men, twelve strong, headed north into the rubber. If the enemy challenged these men they would have one hell-of-a-fight on their hands.

Lima also had listening post duty, so O'Connor and two others went out less than a hundred meters to Lima's front, directly west of the artillery. The terrain to our front was clear, then sloped down to a larger valley, which appeared to be thick jungle. Off in the distance was Cambodia. Coleman handed Lima One's radio to O'Connor so the LP could communicate with Lima Six and Charlie Six.

0030 hours, November 2, 1967. If I was dreaming at all, it must have been about Sydney. Coleman had watch and I was napping in the sanctuary of the sandbags stacked three high extending behind our bunker. "We've got incoming," shouted Coleman. Explosions were loud and frequent; shrapnel ripped everywhere. Coleman climbed into the bunker, as I shook the sleep off.

At this point, O'Connor hauled ass back to the bunker and dove into the sanctuary of the sleeping area. Coleman took the radio back: "I felt naked without it." I suggested O.C. take shelter in the bunker. "No way, I'll stay here." He reported that when the first mortar rounds hit, he was unable to reach Lima Six, so he radioed Charlie Six, who directed that the LP get back to the perimeter.

We kept low, eyes to the front, expecting an assault to follow the mortar attack. Mortar rounds rained on us, Oscar platoon's mortars, and the artillery. Whoever was directing the incoming knew what he was doing. If the enemy followed the mortar barrage with a ground assault, Lima was in trouble with 2nd Squad on ambush.

Sitting in the well of the bunker, I caught a piece of shrapnel in the upper left back. O'Connor was behind me and hollered, "Mac, there's a piece of shrapnel sticking out of your back."

"Well, pull it out, Johnny!" O'Connor did, causing burns on his hand from the hot shrapnel. The wound was a modest one, nothing to worry about. Someone must have called our medic and Doc Houchins showed up.

"Mac, are you hit?"

"Doc, I'm fine. Get back to your bunker. It is too dangerous for you to be out."

The mortar attack slowed. Stuffing two magazines in my pants side pockets and a frag grenade in my shirt I told O'Connor and Coleman that I would check on the CP bunker. Keep a keen eye out to our front as we may get a probe or ground assault at any time.

The sleeping area at the CP bunker had taken a direct hit. John May died instantly. Paul Zima, platoon leader, was seriously wounded. David Estus, the RTO, was hit with multiple shrapnel wounds in the upper and lower back. In the bunker next to the CP, Ken Gardellis had been on guard, sitting on the sandbags. The first round coming in had struck the CP bunker; Ken was hit with shrapnel, knocking him to the ground. Lima's Dennis Beal, 1st Squad team leader, was also wounded during the mortar attack.

I needed to get back to my radio and report to Charlie Six. As I turned from the CP bunker, or what once was a CP bunker, I heard rounds popping from the "ammo pit" where we kept our extra ammunition, grenades, and claymore mines. The pit was round and about four- to five-feet deep. It, too, had taken a direct hit. There was fire in the pit and a couple rounds had "cooked off." This was not good.

Behind the CP bunker, I found a five-gallon water can and dragged it over to the pit, keeping as low as possible. As I tilted the water can to douse the rising flames, rounds cooked off right and left. Seeing the grenades and "thump gun" rounds in the flames cleared my thinking – "I'm damn sure not going to die with this water can in my hands." I moved on.

I checked the Lima bunkers, stating the obvious, finding the guys up and alert, gathering ammo and grenades. The M-60 team, with a new gunner, was ready. Bob Duncan had injured his knee during the battle on October 29th and it swelled double normal size. He couldn't walk much less carry Lima's M-60. Over his objection, he was dusted-off on October 31st. Kimball Myrick, from Mississippi, took over 1st Squad's gun – it was in good hands.

The fire and explosions from the ammo pit lit up the area, so any enemy watching could see us, the bunkers, and the 105s. I prayed they couldn't see that the Lima bunkers were so thinly manned.

Back at my bunker, I told Coleman I needed to talk to Captain Annan. Coleman soon handed me the radio. "Charlie Six, over."

"Charlie Six this is Lima One. Lima Five is KIA, Lima Six is WIA, needs medic, others WIA. I am taking command of the platoon, over." My hope was that my voice did not expose the fear that I felt at that moment.

"Say again, over."

"I am taking command of Lima. Lima Five is KIA, Lima Six is WIA, others also wounded, over"

"[Pause], I read you Lima Charlie. Charlie Six, out."

0100 hours, November 2, 1967. The word went out that Lima Platoon had casualties. Doc Simpson grabbed his medic bag and hustled over to Lima. Arriving at the CP bunker, he first spotted Lima Five. Simpson reached down to attend to Sergeant May, saw that he was dead, and was suddenly blown back aside the bunker and on top of Lieutenant Zima. Dazed and shaken, but otherwise okay, Doc realized that it was the ammo pit that had exploded. He turned his attention to Paul Zima. Lima Six was unable to speak due to wounds to his face and head. Another medic arrived and helped Simpson with Paul Zima and David Estus. Doc advised Charlie Six that Zima required dust-off – he couldn't wait for the morning sun.

Dogface Six couldn't get a medevac bird to come in but a supply chopper did. O'Connor and I ran over to assist the medics. We carried Paul Zima to the chopper as soon as it touched down, placed Zima in and shouted, "Get him out of here!" As we did that, the crew kicked boxes of ammo and grenades off the supply slick. Dogface would need them. In the hurly-burly of the night, we didn't realize how serious Dave Estus' wounds were; he wasn't dusted-off till November 4th.

John May had been very excited because he was to leave the field on November 2nd to head back to Di An and from there to Hawaii. He was meeting his wife on R&R. We never knew whether Mrs. May learned of her husband's death before she left her young children to meet John in Hawaii.

The Ambushes Blow. Monitoring the radio traffic, Coleman advised that the 2nd Squad was reporting movement coming up through the rubber. Suddenly, to the south, the Alpha Company ambush blew their Claymores and headed up a small road inside the rubber that ran straight through the NDP. A Recon platoon bunker on the west edge of the road awaited them, as did an Alpha Company bunker on the east edge. As the approaching dark silhouettes took shape, they hollered, "Ambush coming in!" A soldier standing by the edge of his bunker shouted back, "Ambush, come on in." The Recon grunts began the count as the Alpha ambush reached the NDP: one, two, three, four, five, six, seven, eight, nine, ten, eleven, and twelve. Ambush accounted for – wait. Coming up the road was another silhouette yelling in perfect English, "Ambush coming in." The Recon grunt glanced at his bunker mates; they communicated instantly although not a word was spoken. One grunt hollered back, "Ambush, come on in." When the thirteenth dark shape was within feet of the sentries, he was cut down in a barrage of fire.[8]

Minutes after number thirteen was gunned down, the Delta Company ambush to the east blew their Claymores and headed back to the perimeter.

This left 2nd Squad as the sole ambush out, but not for long. The guys reported that heavy movement was approaching the perimeter from the north and northeast. Charlie Six directed them to blow the ambush and get back to the NDP. The blasting Claymores rocked the brief stillness and seconds later automatic weapons on both sides ripped the night. From my bunker I could see the green tracers coming into and over the Charlie Company line. The 2nd Squad was firing and maneuvering to reach the safety of their bunkers. Mike, November, and Oscar Platoons opened fire on the enemy muzzle flashes. I was horrified with the certainty that the 2nd Squad was caught in the crossfire and could be wiped out. The crossfire was intense, green tracers inbound and red tracers outbound, like ships passing in the night, oblivious to one another, each seeking a target.

The terrain presented a recess that allowed the 2nd Squad to dip below ground level, out of the thunderous crossfire, and work their way over to the road leading south into the Dogface position. Out of breath, adrenalin pumping, Lima Two radioed that the 2nd Squad was back – 12 men present and accounted for. O'Be had monitored my earlier transmission to Charlie Six and knew that Lima had suffered casualties during the mortar attack. Still today, I rejoice at the indescribable sensation of relief, and disbelief, that the 12 emerged from that crossfire and were back with Lima. I told O'Be

---

8 With the morning sun on the horizon, the enemy soldier was visible. He had been struck by an M79 "thump gun" round in his head, caving in the side of his skull. The round had not exploded. I saw the enemy body and asked one of the Recon guys about it: "How could you be so sure that you counted twelve men?"

"I counted twelve", he stated.

"Sure," I said, "but it was dark and confusing. How were you so certain?"

Again, he simply said "I counted twelve, and the other guy coming up shouting 'ambush coming in' was thirteen. He wasn't getting in."

to get his men settled in and ready for action.

"How was your camping trip, over."

"We ran out of marshmallows. You got any coffee on over there, over."

"Negative. We haven't been focused on coffee. I'll make some rot-gut when the sun comes up, over."

"Roger that, Lima Two out."

0200 hours, November 2, 1967. After the three ambush patrols were back with Dogface, there was an eerie silence, akin to being in the "eye" of a hurricane. We all knew something was coming. Then, breaking the silence, over on the east side of the NDP there was a "pop," like a firecracker, then illumination. It was one of our trip flares put out for just this purpose. A second "pop," then a third, a fourth, a fifth merged into a string of firecrackers illuminating everything from the northeast around to the south of the NDP. The advancing enemy soldiers hit the ground and waited for the trip flares to burn out, but Dogface had been warned. The enemy assault was directed at Alpha, Delta, and that portion of Charlie located in the rubber.

Like so many other moments in war throughout history, enemies met on a battlefield where quarter was neither asked for nor given. The intentions of the combatants hadn't changed, only the weapons. Each man present, regardless of uniform, knew to a certainty that this fight was to the finish.

0230 hours, November 2, 1967. On the ground, infantry elements of the 273rd VC Regiment advanced on the Dogface NDP. The mortar platoon and gunships fired flares that illuminated the ground but also cast distorting shadows as they fell among the rubber trees. As the enemy advanced, automatic weapons exchanged fire, grenades exploded, and RPGs whined across the perimeter seeking contact with the camouflaged defensive bunkers interspersed with the rubber trees. Dogface blew Claymores to our

Loc Ninh. 105s positioned in semicircle formed by linkage of Lima and Recon Platoons, loaded with "beehive" rounds during the assault on the Dogface NDP early morning of November 2, 1967.

immediate front as the enemy assault neared the perimeter.

The VC came at Dogface with flamethrowers. The flamethrowers required ignition, which in the night drew the attention of every grunt. When the first flamethrower ignited, every GI within eyeshot shifted their fire on to the guy with the flamethrower. Red tracers erupted from all types of weapons, up and down the line, converging on the soldier with the flamethrower. He and it exploded. A second flamethrower was ignited and just as quickly incinerated.[9]

Oscar platoon, commanded by Lieutenant Emmett Smart from Texas, fired illumination and high explosive rounds at a blistering pace. Fired with minimum charge, the explosive rounds landed within 50 meters outside the NDP. Smart would soon be reassigned to become Lima Six.

At some point during the fight, we heard that the 105s were

9 Later, three more flamethrowers were captured and identified as Soviet models.

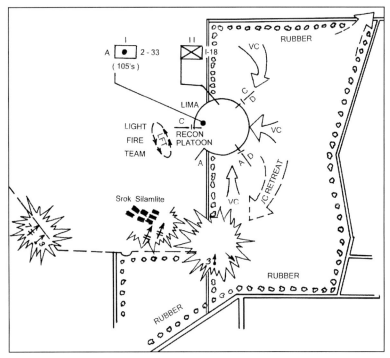

Nov 2, 1967. Battle of Srok Silamlite III.

locked and loaded with "beehive" rounds. The 105mm version contained about 8,000 "flechettes," each about 1-inch long, pointed at one end with small fins on the other, like an arrow. Beehive rounds were to be fired directly into the enemy assault, making the artillery piece into a giant shotgun. If the word came that the artillery guys had to resort to the "beehive" rounds, we grunts would get in our bunkers to avoid being wiped out. This last resort would be taken only if the enemy breached the perimeter and threatened to overrun the artillery unit.

Gunships were overhead directing fire on the advancing enemy infantry. They were joined eventually by Air Force fighter jets. The enemy was ready with twelve 12.7mm anti-aircraft machinegun positions. The 12.7mm heavy machinegun fired .50 caliber rounds at 600 rounds per minute. From our vantage point, we had an unimpeded view of the air battle against the night sky. It was an amazing, frightening, riveting show – red and green tracers from every direction passing each other vertically amid the after-burners of jets being chased by green tracers.

The pilots and door gunners above us were presented with an equally amazing scene – an infantry battalion NDP under attack from several directions; illumination rounds popping overhead, sinking, then replaced; the fireworks display of red and green tracers around the perimeter; explosions of every kind. Add to this the artillery rounds coming in from those NDPs within range and the work of our own mortar platoons, and you had a scene that the men who were there would mark as a once in a lifetime memory.

Map Symbols

⊠   Infantry Unit

▭   Artillery Unit

I   Company, Battery or Troop

II   Battalion or Squadron

X   Brigade

⊣⊢►   50 Cal. Machinegun

O—►   Mortar

◄   Ground Assault Axis of Advance

At one point, framed against the clear black sky to the south of the NDP, I observed an unforgettable scene. The afterburners on the jets allowed us to follow the fighter's path. When the jets dove to strafe or rocket, we could see the ordnance darting to the targets. One flyboy came in firing down at an anti-aircraft position as the 12.7mm fired up at him, colors merging, the rounds seeming to blend together when a midair explosion lit up the sky. Stunned, disbelieving, it appeared that the gooks had just shot down one of our jet fighters. That instant thought exposed just how isolated we were in our patch of nowhere. Then, I saw the jet's afterburner pull up and away, streaking against the black of night. I concluded then, and remain confident, that the jet dropped a bomb which, on the way to the target, was hit by an anti-aircraft round and detonated in flight. The anti-aircraft crew surely cheered their victory and savored the time they bought with their good fortune.[10]

I'm sure that all Dogface soldiers can testify that the interval between midnight and first light seemed a very long time. The 273rd VC Regiment broke contact around 0415 and withdrew to the southeast. For them, this was, indeed, a very, very long night.

Dogface Tested No Longer. We stuck around the NDP for five more days but the 165th NVA Regiment and the 273rd VC Regiment no longer savored the prospect of taking a bite out of Dogface. The battalion had been tested and passed at the head of the class. The enemy had taken on an experienced BRO infantry battalion and suffered the consequences. Initial enemy losses were 220 killed, but after several days of patrols, the final body count reached 263 from the November 2nd battle. The battle is referred to as "Srok Silamlite III." As a testament to the "DePuy" defensive fighting positions, and our artillery, mortar, and air support, Dogface losses in that battle were one killed and eight wounded in action. Lima Platoon of Charlie Company had experienced a disproportionate percentage of those nine casualties.

A day or two after the November 2nd battle, two companies went out on patrol. Lima Charlie got a break – we weren't on the point – but we did go back to that hill where those three enemy soldiers and I had a brief staring contest on November 1st. It was the staging area for the force that hit us shortly after midnight. Throughout the trenches interlacing the hill and surrounding terrain we found their communications wire, ammunition, food, and many other military and personal items.

"The old man was right. This was a trap."

This episode has occupied my mind many times over the years. Although I only served with one infantry battalion in Vietnam, I have read a bit and spoken with many others who paid their dues as combat soldiers in Nam. I believe that many infantry battalion commanders would have sent 1st Squad, maybe all of Lima, up that hill.

They would not have recognized the trap, or they may have seen it as an opportunity to engage the enemy without appreciating that it would be on the enemy's terms.

While on a business trip to Texas many years later, I met General Cavazos, a retired four-star general, for breakfast. Although we had stayed in contact over the years, we had not seen one another since December 1967 in Di An.[11] I reminded him of the call that he had made on November 1st and mused that had he sent Lima up that hill, I wouldn't be here to enjoy our breakfast together. He remembered the call well.

General Cavazos stated that he knew who was on point and was wise enough to pay attention to what his point man reported. When he received my report, the picture, to him, was clear. The enemy's subtle attempt to lure us up that hill signaled a trap and he did not commit his Dogface soldiers unless it was on our terms, not theirs. No argument here.

Major General John Hay, the commanding general of the 1st Infantry Division during the Loc Ninh battles, congratulated the Dogface Battalion and the other units that fought there for our "completely outstanding performance...conducted with power and distinction, contributing immeasurably to victory in Vietnam in the proud name of the Big Red One." At the conclusion of the battles around Loc Ninh, General William Westmoreland, commander of all U.S. forces in Vietnam, congratulated General Hay in praise of the BRO: "This operation is one of the most significant and important that has been conducted in Vietnam, and I am delighted with the tremendous performance of your division. So far as I can see, you

10 After action reports indicate that between 0230 and 0430, airstrikes took out most of the enemy anti-aircraft positions, resulting in 55 known enemy dead.

11 The occasion was Lieutenant Colonel Cavazos's emotional farewell to Charlie Company before he was promoted and given a new assignment.

have just made one mistake, and that is you made it look too easy."

Valorous Unit Citation: Binh Long Province. At that breakfast with General Cavazos, I learned for the first time that the battalion had received the Valorous Unit Award for extraordinary heroism in action during the period October 6 to December 10, 1967. To receive this award, a unit must have performed with marked distinction under difficult and hazardous conditions so as to set it apart from other units participating in the same conflict. The degree of heroism required is equal to that which would warrant the award of the Silver Star, the nation's third highest, to an individual.

The Award's Citation reads as follows:

"The 1st Battalion, 18th Infantry and its assigned units distinguished themselves by extraordinary heroism while engaged in military operations in the Republic of Vietnam during the period 6 October to 10 December 1967. During this period the Battalion engaged in eight significant battles and several smaller skirmishes with hostile forces in the Binh Long Province. Because of the determined fighting spirit of the Battalion and its superior tactical capabilities, these engagements resulted in severe losses of both men and equipment by the North Vietnamese Army and Viet Cong forces while the friendly forces suffered only minor losses. Operation Shenandoah II, designed to clear the northern area of hostile operations resulted in approximately 60 days and nights of almost continuous action and daily contacts. At the conclusion of the period, Highway 13 south of Quan Loi was opened and secured. The Battalion's excellence in both offensive and defensive operations was undeniably proven during this period. Its expert leadership and extraordinary combat effectiveness were responsible for gaining unquestionable superiority over enemy forces in the area, resulting in over 700 enemy killed in action and the capture of numerous weapons and much equipment. The men of the 1st Battalion, 18th Infantry, 1st Division, displayed extraordinary heroism and devotion to duty which are in keeping with the highest traditions of the military service and reflect great credit upon themselves, the 1st Infantry Division, and the United States Army."

This story, a grunt's eye view, describes a piece of a portion of the 66-day period during which the Dogface Battalion "displayed extraordinary heroism and devotion to duty." I'll leave it to others to tell the rest of the stories.

After Thoughts. Throughout the generations, our warriors have never been able to pick their wars. They serve when and where "we the people" tell them. The Vietnam War was no exception. When the World War II generation sent us off to fight in Vietnam, and by war's end, collectively, we looked something like this:

• About 2.7 million men and women in the U.S. military served in the Vietnam "war zone."

• Vietnam veterans represented about 9.7 percent of their generation.

• There are 58,267 names on "The Wall," and 304,000 others were wounded (this does not count "Agent Orange" casualties, those who died later of wounds received in the war, or those who survived but never really made it back from that war).

• Of the Vietnam War dead, 25,493 were 20 years of age or younger (43.8% of war dead).

• Draftees comprised 25% of total forces in country, and accounted for 30.4% of Vietnam War dead, 17,672 men.

• Just less than 7,500 women served in Vietnam, of whom 83.5 percent were nurses.

Truth be told, Vietnam is the defining event for men of my generation, the so-called "baby boomers." There were options for men of the Vietnam generation, and I opted to volunteer for the draft. Not once have I regretted that I served when and where "the people" told me to go. It was my turn to pay dues. Having said that, those who sent America to fight in Vietnam, and kept us there till 1975, have much for which they should answer. As always, history will render the verdict.

I'm compelled to touch on one last point. When people here and abroad discuss the Vietnam War, it is often touted as a military defeat for the United States. This suggests that our ground forces were beaten by the ground forces of North Vietnam and the Viet Cong. I respect the enemy we fought. They were brave and tenacious fighters. My story certifies that the enemy we faced was a worthy adversary.

I'm reminded, though, of a conversation described by Harry Summers, Jr. Colonel (Ret), and a former Big Red One officer, in his acclaimed book *On Strategy, – A Critical Analysis of the Vietnam War.* Summers describes a conversation in Hanoi on April 25, 1975 between himself as Chief, Negotiations Division, U.S. Delegation, Four Party Joint Military Team, and a Colonel Tu, Chief of the North Vietnamese Delegation. The conversation went like this:

"You know you never defeated us on the battlefield," said the American colonel.

The North Vietnamese colonel pondered this remark a moment.

"That may be so," he replied, "but it is also irrelevant."

It is not irrelevant to me, a soldier who fought in some of those battles.

Ambush at Loc Ninh

David Gilbert

November 1, 1967 and the early morning of November 2nd stick in my mind more than any other time. I was still a FNG. Second Squad, Lima Platoon, had ambush, and this was a real bad area to be going on an ambush. Loc Ninh was a hotspot, and there were plenty of NVA/VC looking to cause trouble. The Loc Ninh special forces camp and airstrip was an enemy target, and we were placed between Loc Ninh and the Cambodian border.

As daylight ended, our ambush went out from the NDP. We set up in a good spot, but it was spooky. Sergeant Bob O'Brien (O'Be) was our squad leader and a damn good one. We also had Ray Etherton (Porky) and a few other old timers; so we were in good shape as far as manpower. Porky had been in the field for a while, and I felt safe with him as my team leader.

We set our Claymores out so they would cover the area in order to do the most damage. We were alert and awake just waiting for something to happen. O'Be said if he caught anybody falling asleep, he would kick their ass. My eyes were as open as they would go.

Shortly after midnight the NDP was hit by mortars, and the attack seemed to last forever. After the attack, we had movement coming our way through the rubber trees. At some point I heard a chopper land near where Lima Platoon was positioned; this must have meant some guys got hit. Then we heard Claymores blow south of us and again over to the east side of the NDP. This left our squad as the sole remaining ambush still out. The RTO reported to Charlie Six that enemy movement was increasing and closing in on us and the NDP. Charlie Six directed that we blow the ambush and get back to the perimeter.

O'Be and Porky were low-crawling from one guy to another to let us know to be ready to blow the ambush. I was nervous. By the time Porky and O'Be made the rounds, we could hear the enemy getting closer. A few days earlier I received my real baptism in a fight when the enemy was on top of us, but still I worried about my overall inexperience. I was shaking and hoped that no one noticed.

O'Be told us to blow our Claymores and Charlie started firing AK-47s in all directions, unsure where we were located. Porky didn't think they knew where we were positioned; so he wanted us to make our getaway as quiet as possible. Of course, when we blew the Claymores, we had no choice but to haul ass out of the ambush site. An infantry squad makes a lot of noise when moving. As we headed out, O'Be stood up and started firing back. Porky just about went mad and started yelling at O'Be to forget the firefight and get back to the NDP. As we were running, the enemy soldiers were behind us, still firing. All we could hear were those AK-47s being fired at us. We were also worried about the guys at the NDP thinking we were unfriendlies and shooting at us. All manner of things were going through my mind.

That is precisely what happened. As the enemy fired at us, the men on the perimeter, Mike, November, and Oscar Platoons let go on them. Lucky for us we found a draw that let us move west back to the small road that ran through the NDP, which kept us out of the crossfire. The firing was intense, tracers zipping and crisscrossing, red and green rounds of death. It seemed to suck the oxygen out of the air.

I was glad not to be in the rear of the ambush coming back in. The last guy had it the worst. Trying to run in the dark, not lose contact with the guys in front, and still maintaining the rearguard for enemy soldiers attempting to follow Second Squad back to the NDP was a dangerous job. I wish I could remember who did that job – I'd like to thank him – something we all were too busy to think about back then.

O'Brien and the old timers did a great job with us FNGs, and Second Squad maneuvered as one. The AK-47 rounds hit all around us, close, very close. It is still nearly impossible to believe that 12 grunts made it through that gauntlet without taking casualties.

As we neared the NDP, our mortar platoon fired flares so we could see our way back in. Problem was that the enemy could see what we did. We had a certain spot that needed to be found to get back in to the NDP, and we were looking to hit it first time around. Oscar Platoon was now dropping the kill rounds right in the bad boys' laps. The mortar rounds were doing beau coup damage and creating distance between us and them. Trip flares were going off as we entered the perimeter, and periodic explosions were popping off from Lima Platoon's extra ammo pit. The pit was behind the command bunker, and I was able to see that the command bunker had taken a direct hit. Maybe this was the reason for the chopper we heard while on ambush. When Second Squad made it to our squad bunkers, I experienced the greatest relief of my relatively short life.

Even though it has been over 40 years, I thank all the guys who had anything to do with us making it back safe, and the guys back at the bunkers that were waiting for us and sweating it out. If we had taken casualties, I know that First Squad under Patrick McLaughlin, (Sergeant Mac) was ready and willing to come out and get us back to the NDP. That's what made Charlie Company and Lima Platoon so good; we were always ready to take care of our own. In the end, we fought and cared for each other. Tommy Mercer and Kenny Gardellis said it looked like death was chasing us back in. Our eyes were wide open, sucking air, some guys yelling, and others crying for joy that 12 came back alive.

Back at the platoon bunkers our joy was diminished when we learned that John May was killed. He was our platoon sergeant, and we all respected and liked him. And the Huey we heard come in

Ernest (Smokey) McNeill

while we were still out was to dust-off Lieutenant Paul Zima, Lima's platoon leader. Zima was a good platoon leader, and we all prayed that he would make it. Several other Lima grunts were hit by shrapnel during the mortar attack but did not require dust-off. Mac took command of Lima Platoon, and I certainly felt safe with him. Back in our bunkers we readied for the attack that the Dogface Battalion knew was coming. The attack came, but the sun rose in the morning – at least for us.

Smokey McNeill

Smokey McNeill

I, Ernest (Smokey) McNeill, served in the Vietnam War from July 67 to July 68. I was sent to the 1st Division's Charlie Company, 1st Battalion, 18th Infantry Regiment. I served as a Kilo for the squad leader. I carried an M-16 rifle. I became a man over there. The first firefight I was in, they awarded me a Bronze Star medal. At that time I was serving as a grenadier on an ambush patrol near Lai Khe.

As the unit was moving into position, a large Viet Cong force employing grenade launchers, small arms, and automatic weapons opened fire on us. Several men were wounded in the initial barrage. Because of my quick thinking, I exposed myself to the hostile fire and moved to a forward position, and I placed suppressive fire onto the insurgents, which enabled the medical aid men to reach the casualties.

Contact!

Bob Duncan

I t was October 29, 1967, 12:09 hours – our "High Noon." As Lima Platoon's M-60 machine gunner, the first of two gun-teams in 1st Platoon Charlie Company of the 1st Battalion, 18th Infantry Regiment of the Army's famous, 1st Infantry Division, I had been in-country nine months of my twelve-month tour of duty – a couple of weeks into Operation Shenandoah II.

Dogface Charlie, First Platoon, First Squad Takes It to Them. Under direct command of the Swamp Rat's Battalion Commander, Lieutenant Colonel Cavazos, radio call sign Dogface Six; Charlie Company Commander Captain Bill Annan, radio call sign Dogface Charlie Six; and our platoon leader, Lieutenant Paul Zima, call sign Lima Six; we were part of the first lift to set down as the remainder of the Dogface Battalion was air assaulted in via helicopters. First Platoon of Charlie Company had been informed that we had point. Upper command must have liked what Charlie

Company did and how we did it. The enemy was reportedly half a click away, at the edge of the rubber.

We approached the bad guys on an oblique angle in an inverted V-shaped formation. Our patrol's First Squad point man, O.C., made first contact. He yelled, "Gooks." I believe Mac (1st Squad leader, Sergeant Patrick McLaughlin) then made contact with part of a large forward enemy element well entrenched and hidden in dry irrigation trenches that interlaced the huge rubber tree plantation. The bad guys were in our faces and outnumbered us. My gun team was about five soldiers back behind O.C. when the shooting started; we ended up next to him as he tried to direct everyone's attention to the enemy who were but feet away. He immediately threw hand grenades into the trenches as we could see their heads and bayonets scurrying back and forth, then disappearing in short blurs and pieces of body matter blown out of the trenches. Mac had his own face-to-face shootout with an enemy soldier as the first squad moved up just feet away from the front trenches. Mac killed him.

O.C. – Point-man John O'Connor. O.C. shouted, "Gooks! They're in the trenches. They're in the trenches!" All of this happened in only seconds. First Squad maneuvered into an on line formation, closing with the enemy, and taking cover just feet from them face-to-face. We followed O.C.'s lead and tossed a lot of grenades. With a four-second fuse, pulling the pins, and popping the spoons on a very quick one-two count prevented the enemy from throwing them back at us.

As our squad and Second Squad tried to squeeze closer to O.C., we'd move or toss the remainder of our pinned (unarmed) grenades forward to O.C. and those of us preparing to kill the enemy in the front trench. One went just past us and rolled around a rubber tree mound only to end up right next to our platoon leader, Zima, and his RTO, Charlie Lima "Kilo." Lieutenant Zima was trying to give a radio size-up as all this was going down so rapidly. Zima was therefore on the horn (handset to his ear) when he saw the grenade coming his way. He wasn't letting go of that phone. However, his poor Kilo still had the radio on his back. As our lieutenant evaded our inert grenade, this resulted in the Kilo being dragged a few feet while our lieutenant tried to get away from our grenade. In the middle of this life-and-death fight, O.C. and a few of us actually found this funny. Stranger things have happened in combat, but that was funny.

As point, O.C.'s vigilance, fast reactions, and nerve saved a lot of lives that day. I don't think our hardcore enemy knew what hit them. The trenches and our suppressive fire would only allow them lateral movement. We took full advantage of their dilemma.

Fire Team Work and Sacrifice. Half of a smoking, spinning enemy body flipped up and out of the trench directly in front of us. First Squad stood in unison and overran the enemy, killing all in the forward trench, and jumping on top of them because we still needed cover as more enemy continued to unload on us. I can remember trying to sweep the entrenched enemy as we stood to hit them but only moving my fire partially up the trench for fear of hitting one of our own guys jumping down into it. By now our efforts had paid off as the enemy had more or less piled up, clogging their egress, forcing them to move against or away from us as we hit them fast and hard in those trenches. There was some hand-to-hand. It didn't last long, but long enough for me to engage an enemy soldier.

At this moment, as our guys were pouring into the fight, a lot of attention was still focused out to 100 feet or so. We could see the movement out there, but for seconds we still didn't know that O.C. had seen the enemy just feet away from us. I had placed two rounds into a soldier in front of me as he took my feet out from under me from his lower trench position. My left knee and hip took the hit as I landed on his head, incurring a significant knee and hip injury. The two of us were pinched together between the trench's walls. Eye to eye we grappled, my sizzling machine-gun's barrel had been turned around and was now facing directly at me, when I leaned forward and depressed the 60's trigger with my thumb, it jumped after a few last rounds were fed into the 60's chamber. My M-60 machine gun's hot barrel left a lifetime tattoo on my upper arm – a red-hot branding and constant reminder of how lucky combat soldiers could be. As providence would have it, the soldier I had already fired on seconds earlier fought with the energy of a rag doll. It didn't take much to kill him.

Mac took out a four-man enemy machine gun team as First Squad, 1st Platoon had taken the trench with more to come. With the exception of the face-to-face encounters, it was hard to know exactly whose bullets and grenades killed which enemy-soldiers in such a target-rich environment.

Fire-Team Leader John Willett miraculously survived thanks to John O'Connor (O.C.), our brave medic "Doc" Simpson, and Sergeant Mac who lifted him to a stretcher while still under sporadic gunfire. The cracking of enemy gun and rifle fire were still in the air and not subsiding, and some AK fire still coming down our captured bloody trench. It was apparent we would not be in those trenches for long if we were to keep the momentum rolling in our favor.

That familiar "cracking and humming" was the sound of "incoming" exceeding the sound barrier near your ear…enemy bullets were very close…too close.

O'Be (2nd Squad Leader, Sergeant Bob O'Brien) immediately brought his squad up behind us in a classic fire maneuver method. Having a bigger picture, he killed the enemy soldier(s) who had flanked us. O'Be wasn't silent about his actions. I heard more than a few expletives. He saved a lot of lives that day as well, making sure that we checked for enemy in the trees. When in doubt, recon by fire would get a tree-sniper's attention, causing him to move. If we could see them, we could kill them before they killed us – the soon-

Lai Khe. 1968. Biser, Tom Mercer

er, the better in such close-in fighting.

Persistent enemy rifle fire was still coming in – not a good sign. All I had time to do was double check for wounded and KIA, reload, then verify my team's ammo and manpower count: PFCs Myrick and Biser still effective. There was no time to say goodbye as Willett was hurried to a medevac chopper for dust-off. Our guys with Doc Simpson were doing what they could do for Johnny Willett.

For a very short amount of time, what was left of my shocked and stunned gun team (minus those of our squad who had carried Sergeant Willett) were not with me as we were ordered by the lieutenant to advance out of those trenches. We had not only won that first round with dedication and sacrifice, but let the enemy know that we were not simply average soldiers – we had attitudes.

It wasn't long before those who had carried Willett to the dust-off would return with this same attitude. Our squad and gun-team led the way, placing suppressive fire out ahead of us as the rest of Charlie Company and the Dogface Battalion joined us in engaging and routing the enemy out of the many irrigation trenches that honeycombed the rubber.

With heavy enemy movement in many directions, our platoon leader, Lieutenant Zima, ordered us to move up on line and make a hard right to engage more entrenched enemy. We were moving and shooting at a steady pace, bringing hell into that rubber tree grove while putting direct and suppressive fire out ahead of us, consequently, giving the remainder of Charlie Company and the Dogface Battalion time to catch up to us in the many shadows of the trees. My left knee had grown to about the size of a football. I felt nothing but anger as the adrenaline rush pushed my 20-year-old body forward. I engaged the enemy wherever they were. The 60 was working. However, belted .308 NATO machine-gun ammo was quickly becoming harder to find. My assistant gunners, Privates First Class Myrick and Biser were doing their best just to stay alive. This left me scrounging where I could while engaging more enemy.

The On-line Sweep. I'm reminded of our Dogface Battalion formation when I see Civil War reenactments of foot soldiers charging their opposition on-line in a Napoleonic face-to-face maneuver. Until that day, I'd never been part of such a display of direct and close-in ground fire at such short distances – exercises in the formation of raw and unyielding firepower.

Lieutenant Zima and Sergeant May were shouting directions and pointing. Our newer platoon sergeant didn't fully understand that my assistant gunner had to be closer to me than the normal spread of our riflemen. As a crew-served weapon, the "gun" had to be fed more ammo than a rifle to maintain effectiveness. We pushed forward online; the noise became deafening. We stepped over and around enemy dead, clearing trenches, and shouting directions and locations to one another. The irrigation trenches may have offered the enemy protection, but this also provided us the advantage of knowing where to find and kill them before they could surprise us. It wouldn't be long before they realized it was not a good time to fight us. They chose to uproot and run like hell.

My knee-pain worsened, but I couldn't afford to fall back because it would cause a break in our 150-200 meter battle line. Falling behind was not an option. It narrowed my M-60's field of fire, thus creating an unsafe friendly firing situation, virtually rendering the gun's effectiveness to zero. I prayed to God for the physical strength to stay online. By now, large fragmented groups of armed unfriendlies were running away from us at distances of 25 to 250 meters. Leaning against some of the rubber trees to stabilize my shoulder shots, the 60's sights needed minor adjustments to take one, two, or three enemy out with one trigger squeeze. "Layin' it down." "Bringing p°°s on 'em." Such targets of opportunity were rare.

Aim small and tight and hit small and tight. The sounds of small arms bullets yawing instead of moving through the air with greater accuracy. This humming sound meant that wounds sustained at such close range would be more devastating as well. As long as we kept moving on them, we'd kill them any way we could. We took what we could get while targets of opportunity still existed. The many large rubber trees also offered hard cover and helped stabilize us in our firing positions – field expediency.

Finally Some Artillery for Us. Our FOs were the best. Battlefield conditions were always changing. One of my sight pictures was right on the money. I had a more concentrated group of enemy almost in the open. As I proceeded to squeeze off, a burst of artillery came through the treetops and hit the trees on a projected angle from my right, and the group that I had focused on disappeared in the blink of an eye. There was a red dust cloud, and they were gone. I was impressed and felt empowered. Army choppers had dropped a wooden case (800 rounds) of belted M-60 ammo through the trees. More than happy to lessen their loads, riflemen were throwing the 60 ammo cans of 200 rounds at me when they could.

Decorations and Awards. Most of us had just come off of this sweep as the enemy, still in force, continued to make its presence known. After killing even more hardcore enemy that afternoon, many of us were lucky to give witness to our division commanding general's personal pinning of the Silver Star on O.C. (Specialist John O'Connor). I received several of my awards a decade later to include another Bronze Star with V Device First Oak Leaf Cluster and a Purple Heart. A very special Valorous Unit Citation was issued 35 years later to all Dogface combatants involved in Operation Shenandoah II. To this day my most cherished award is the U.S. Army's CIB (Combat Infantry Badge), which I had earned six or seven months prior to this contact.

Appreciation. I'd be remiss without mentioning my original assistant gunner, Jack Volker. He had a big heart and fed several of us his extra water throughout his tour of duty. After Jack Volker was reassigned as 2nd Gun Team Leader, Private First Class Myrick courageously filled in Jack's spot for me. Myrick was a great soldier and with me all the way. I couldn't have had a better soldier on my team. My other ammo bearer, Private First Class Daniel Biser, was with us on our first contact of October 29 but didn't stay long due to malaria.

Decades Later. In 2002, about 40 miles outside Phoenix with about 500 miles to home, my beautiful wife of 28 years died of heart failure. She was 55. Somehow she inherently knew more about the grunt's role in America's Vietnam War than any noncombatant I've ever known. Medically, she had no heart history. My CPR was good; however, the Lord needed her more than I did. We were just finishing a 7,000 mile road-trip that included a visit with my comrade-in-arms, John Willett, and his wife, in New Hampshire enjoying the fall colors. John had stopped a lot of bullets for our squad; a price he and his wife have paid for the rest of their lives.

Over many happy years, she had become my rock, subliminally replacing Charlie Company with her love, courage, and com-

Left to right: Joe Rosencranse, John Willett, John Blackburn?, unknown, Bob (Dunk) Duncan, Jack Volker aka "Voka."

radeship. As good as American soldiers are, too often luck or God has more to do with living or dying than good soldiering. Mary had somehow become a great soldier, a wonderful example of how to appreciate our military's tremendous sacrifice, especially her love of God and country. An even greater example of how to love and live life. As a retired professional firefighter, I was well practiced at CPR; I lost her nonetheless. She did regain consciousness long enough for me to tell her that I loved her. Alone in a drizzly mist of rain, I recall looking up at a huge, fluttering American flag on the hospital's roof. It was waving heavily and slowly recovering in the wind. So, I sat there in the grease of a parking lot with the room she had passed away in just feet from what felt like a bad dream, waiting for the strength of God, Charlie Company and/or Mary to miraculously return and tell me her passing was a bad joke. Nothing was real, but her passing and Nam sure as hell was.

Machine Gunner

Bob Duncan

I never guessed that I'd end up carrying a machine gun. Not too many grunts volunteered to carry the M-60. The two-gun per platoon limit placed a lot of responsibility on the M-60 gunner and his team. There was constant mental pressure on gunners to keep their weapons operational. Gunners must have an aggressive mindset. I trained my team so that every one of them knew how to clean and operate our 60. Like a magnet, it draws enemy fire as well. I was trained to fire in groups of 6-9 rounds per trigger squeeze. The cyclic (continuous) rate of fire was 550 rounds per minute, but as a sustained rate that could heat the barrel up, causing a "cook-off" and/or blow the hardened rifling out of the barrel, which would eventually cause a stoppage that could allow the enemy to ruin our day. It requires more time to clean and maintain and camouflage. It can be very cumbersome to move through thick brush or bamboo. Unlike rifles, we carried the heavy M-60 in different ways. Too many of my superiors forgot that it is a "crew-served" weapon – our crews were often low on manpower. Consequently, carrying the required 1,200-1,500 belted rounds was a constant hassle, and at times created some animosity between myself and those poor riflemen who in turn had to hump an ammo can of 200 rounds. Gunners also carried a .45 caliber Colt pistol…something more to clean.

Memories of Vietnam

J. David "Doc" Simpson Jr.

As a combat medic in Vietnam, I have many memories. I had only been in Vietnam a few months. I don't remember where, but it was late at night, and we came under mortar and small arms attack. The weather was raining and had turned to pure soup. It wasn't long into the engagement when I heard the dreaded word "Doc." When I got to the location, there was a young black man wounded and losing blood really fast. We got him under a poncho, where I began working on him trying to get the bleeding stopped. In the meantime, a radio call went out to get a dust-off in as quickly as possible. The weather was just too bad, and they radioed back that they could not get in. The man I was caring for asked me if he was going to die. I tried to the best of my ability to assure him he was not, but in my mind I knew if we didn't get him out real soon, he would not have a chance. As we waited for the medevac, it was obvious that time was running out. He was deteriorating fast and started calling out for his mother. I'm sure if he could have had her by his side, he believed he would be all right. As I held him in my arms, his last words were "Mama," "Mama," "Mama," and he was gone. He was the first one I lost, and all I could do was hold him and cry. It still haunts me, and I still cry.

Then came Loc Ninh. We had just touched down. A perimeter was established. I was sitting there cleaning my M-16 when I heard, "Saddle up, Doc. We are going on patrol." My M-16 was one of the stainless-steel type, noted for jamming. I always carried a .45 and an M-16. For some reason that gave me some sort of comfort to have a backup. As we left on patrol, we had only gone a short distance out when point man yelled, "Gooks," and it seemed like all hell broke loose. Sergeant McLaughlin cried out, "Doc," and I headed in the direction of his call. When I got there, John Willett was lying in one of the drainage ditches in the rubber plantation. He was shot in the head and brain matter was coming out, and Sergeant Mac had stopped the bleeding as much as possible by sticking his finger in the bullet hole. Sergeant Mac turned Willett over to me, and I wrapped his head with a large bandage, trying to be careful with his brains. While I was getting ready to start an IV, a Viet Cong popped up from another ditch nearby and was ready to shoot me. He didn't have a chance to do that because of Johnny O'Connor, Pat McLaughlin, Kenny Gardellis, and Tom Mercer. They all must have shot him at the same time, and Johnny threw a hand grenade in the ditch. It blew him out of the hole like a rag doll. We continued to fight until we had routed the enemy and made our way back inside the perimeter. That was the longest battle I was in while in Vietnam. It lasted several days, and at times I wondered if we would get out alive. The fighting was very intense with small arms, machine guns, mortars, and even flamethrowers. It got so bad one time, as I sat in a bunker, I wrote what I thought would be my last letter. Later I thought to myself, "Who's going to mail this?" The second night our ammo dump was hit by an RPG, and rounds and hand grenades started going off. A sergeant was hit and set on fire. I tried to pull him out, but something exploded and blew me back about ten feet into a sump hole we had dug. The sergeant was dead, but I had landed beside a lieutenant who was hit in his mouth. I began to treat him as things continued to explode around us. By morning things had settled down, and we were able to evaluate the situation. Many of the Viet Cong laid dead everywhere. Hundreds of them lying in the sun began to swell pretty fast because of the 115 degree days. Just a matter of hours had passed, and they began to smell, and their size could be compared to a large pig. Some of the ones that were captured told us that they had been given orders to overrun us at any cost. We had engaged the 273rd NVA Regiment, but the 1st ID, 1/18th Infantry held our ground. The sergeant that was killed at the ammo dump still hangs with me. He and I had just talked the day before, and he told me he was going to Hawaii to meet his wife. He was to take a chopper out the next day when supplies came in. He seemed to be so excited to be going on R & R and with her. I often think of him and his wife. She was probably on her way to Hawaii and had to be told when she landed.

These are just a few memories, but a book could be written about the life of a combat medic and the decisions that had to be made in a matter of seconds.

Journey through the RVN with the Big Red One

Tom Pippin

I arrived in-country in October of 1967 and was assigned as platoon leader of Mike Platoon in Charlie Company. We did not see much action while I was the platoon leader. In fact, it was boring since the company had the responsibility of guarding Highway 13 much of the time. This was done in NDPs and included patrols and some ambush patrols. What an awesome site and smell! Highway 13 was the major road out of Di An and Lai Khe. I had the good fortune of being under the tutelage of Staff Sergeant Kia as my platoon sergeant – Sergeant Kia taught me how to deal with many problems and was protective of all of us. He survived the war and resided in Hawaii. We have recently found out that he has passed away.

I was promoted to company XO in mid-January. The Tet Offensive hit us on January 31, 1968 while I was in Lai Khe prepar-

Lieutenant Tom Pippin coming to pay the guys of C Company.

ing to go forward with the pay for the troops. This was the closest that we would ever get to what it was like during World War II with the shelling from the Germans. What a night! I was able to get to Saigon where the company had responsibility for the power plant of Saigon. From what I remember, we lost one of ours there in Saigon – Granadas.

In late March 1968, I was promoted to assistant adjutant of the battalion and Lieutenant Emmitt Smart was promoted to the XO. May of 1968 I was promoted to General Ware's staff in Lai Khe under the Chief of Staff (Archie!). My position was liaison officer to USARV – I made two trips daily to Headquarters USARV in Saigon via helicopter. I was responsible for delivering all correspondence to USARV from 1st Division Headquarters. I also kept the maps up-to-date in the Situation Room for USARV. I was in this position until my trip home in November 1968 – newly promoted to captain, and I was assigned to Ft. Ord, Califronia, for a year – and then out of the Army.

While serving as Staff Officer under General Keith Ware, I saw him as one great leader – a soldier's general. He would bring hell down on Charlie if the division was under attack or attacking. His pilots were my hoochmates in Lai Khe – Captain Gerald Plunkett and Chief Warrant Officer 2 Bill Manzanares. On September 13th, 1968, General Ware and all on board his chopper were shot down by NV Regulars. Not only was General Ware lost, but the G3, aide,

Command Sergeant Major Joseph Venable, pilots, crew chief and door gunner were lost. A sad day for the 1st Division and the US Army. General Ware was a Medal of Honor recipient from WWII. My wife and I visited the RVN Wall in DC years ago and paid our respects for all of these members of the First. We then went over to Arlington and found General Ware's grave – in the middle of all of the grunts of the Vietnam War – with us and not with the dignitaries!

Unusual Meeting

Bob Quimby

I was assigned to Charlie Company-November Platoon-1/18, and it would be my new home for the next year. All the guys seemed to keep away from all of us FNGs at first, and I didn't know why. I would find out later. The second day in the field my emotions got the best of me. My stomach was really going crazy. Come to find out I was nervous, and it was causing my body to go all to hell.

The company was dug in a safe area, but still I was scared to death and could not sleep at all. Everyone else took advantage of the easy time and got plenty of sleep when they were not on guard duty. All I could think about was the dinner the night before and how sick it made me: sliced cucumbers and dressing.

After a few months of being in the field, I was getting used to the things we had to endure each day. Even though I didn't like it, there was nothing I could do except finish my tour and hope I made it home. I was now assigned to be November 6 Kilo, at first I didn't like the sound of it, but after a while I got use to it.

The platoon leader was one of the toughest and most competent men I had ever met. He was always concerned for the health and safety of his men and was right there with them in all firefights. Our first major contact with the VC came in October, 1967. Up until then we were just involved in small firefights, but nothing big – yet! But as anybody who fought in Vietnam knows, even a small fight can be deadly. You had to stay alert at all times.

Finally the day came when I lost a good friend, Jim Dossett, while he was on point. He was killed in action on October 5, 1967. Then I realized I was really in a war zone. Jim will be in my memory forever.

While at Loc Ninh I got real sick with the worst fever I have ever had. I thought I was going to die. Finally I went to see the medic. The next morning I still wasn't feeling well. My body wasn't working right, and my temperature was really up there. The medic checked me out and told me I was going back to Lai Khe to get checked out by a doctor. I was just about totally delirious.

I remember knocking on a door after noticing a red cross on the roof. A staff member let me in, and I remember someone asking me if I had been under a lot of stress. I answered by saying I had just come in from Loc Ninh. I was told by the doctor I had high-temp/stress/dehydration. I didn't know what it was, but I never want to have it again. My fever finally broke after about eight days, and they informed me I was ready to go back out to the field.

In 1968, Charlie Company was in all kinds of battles and a lot of bad firefights. We were like anybody who was in Vietnam; we got hit a lot while on patrol and ran into a few ambushes, but we always seemed to come out better than the VC did. That's when our good leadership paid off. We were a battle-tested unit and knew how to fight when the time came.

I served another few weeks as November 6 Kilo and then was assigned to Charlie Six, for Captain Bill Annan. He was a very competent person and a great commander. Soon Captain Annan left and Captain Phil McClure took over in January 1968, and again we had

1968. Nov. Platoon mixed with headquarters RTOs, left to right: Buffalo Boy, unknown, Bob Quimby, Tony J. Kulikowski Jr., back off patrol.

another great company commander. He watched out for his men and was always there when the action started. Captain McClure's first tour was with the South Vietnamese troops, so he was use to fighting. We were a great team, and we went through the 1968 Tet Offensive and other fun stuff. Captain McClure was a caring person, but what we call a real bad ass when he needed to be, but he was always good to me. After Captain McClure left, Captain Sam Downing took over as Charlie Six, and I didn't have to carry the 26-lb. radio anymore; I was getting short. It was so nice to go on patrol without that heavy load on my back.

On Veterans Day 2003 while at the Vietnam Veterans Memorial and walking by the nurse's statue, I noticed a Vet standing there with his wife. He had a 1st Division patch on and the years 1967-1968 under it. We started talking and come to find out he was the medic that treated me at Lai Khe and the same person who opened the door and let me in. For some reason he remembered me from that day; I guess I was that messed up. His wife confirmed that he had talked about a vet he had treated during his tour in Vietnam. I was never going to tell anybody about this because it would be so hard for anybody to believe. We have gotten together a few times, and I even went to one of their son's weddings.

In the intervening years I have dealt with the damage to my spirit and mind. I have received a great deal of help, sometimes contentiously, from the VA, and was able to return to the country of Vietnam with a caring group of people known as Veterans of Vietnam Restoration Project (www.vvrp.org). It was a significant step in my return to health. Also I enjoy some rewarding relationships with other vets, some who gave orders and some who followed orders. One thing is for sure, I will never forget the men I served with in Vietnam.

Highway 13 – Thunder Road

Kenny Gardellis

Thanksgiving 1967. At one point during my tour we helped keep Highway 13 open for transport and commerce. Minesweeping is done on Highway 13 every day, and we patrol and guard the special mine sweep teams that go up the road with handheld metal detectors – a convoy of trucks and Vietnamese traffic following. We sit off the sides of the road in the woods in groups of two, spread out for miles, all day to prevent any VC from mining the road after we sweep it. We don't encounter any trouble while we are in the woods. The Vietnamese civilians often find us and sell us Coca-Cola, French bread, and sunglasses. When the young dudes on motorbikes find out we are going to be guarding all day, they take orders for food, pot, prostitutes, whiskey, a sandwich, etc. They go to the nearest village or Saigon and come back with a woman on the back of the bike plus other supplies that Americans love. Most guys

Some hard-core Charlie dudes off Highway 13 ("Thunder Road"). 1967.

don't buy and don't particularly want a motorbike riding up to them just in case the VC are around, but there are always some guys who do leave their posts for a daytime fling with the pretty backseat riders. The more intense times are in the mornings walking behind the minesweepers to protect them in case of a VC ambush. Some days I am within 20 to 30 feet of the minesweepers. If a mine explodes, it will kill the guy sweeping, and, depending how big the blast (presumably pretty big since they were placed there to blow up trucks), it would probably kill or injure us as well. I don't recall ever finding any and none blew up, but it was tense and tedious work.

We actually have our Thanksgiving dinner in an NDP near Highway 13. The Army flies in some turkey, mashed potatoes, gravy, cranberry sauce, and cold soda to our NDP. It is great, and the food is surprisingly tasty, hot, and fresh. It sure beats eating cold C-rations out of the cans.

The Capture of Ho Chi Minh

Michael A. Shapiro

October 1967 to October 1968. We were in an NDP out on Thunder Road: a long, flat stretch of red clay dirt that ran through the jungle for miles in both directions. Five miles to the north was a small Vietnamese village and eight miles the other way there was another village, and every day the villagers from the north would travel on their bicycles and mopeds to the southern village to sell their chickens and vegetables. During our deployment there I had an exchange of words with my platoon leader, the result of which was that I was put on the road closing detail for the week we were there. At first light in the morning and again at dusk, two brothers and I went out to the road and pulled barbwire rolls across it to close it to traffic for the night or re-open it again in the morning. On the second night of the detail, we had dragged the three rolls of wire across the road and were starting back to the NDP when I noticed a Vietnamese man coming down the road on his bicycle. The Vietnamese knew that they had to be back in their village by nightfall; if they weren't, they could be shot because anything moving after dark was presumed to be VC. Anyway, this old Vietnamese man was peddling his butt off – head down, trucking hard for home, knowing that it was getting dark and he had miscalculated. We saw that he wasn't slowing down and I said to my friends, "I don't think he sees the barbed wire." We stopped walking and watched him and could see that he was going to go full speed into the wire; so I waved my arms and called out, "Hey, Hey, HEY!" But the guy on the bike didn't pay any attention to me, and he rode right into the barbwire rolls, flipped over the handle bars, and lay entangled in the wire.

We laughed, of course, because it was kind of like seeing the

February 1968. Michael Shapiro in middle.

keystone cops, but we walked back out to the road and while my two buddies searched his satchels for grenades and weapons, I started to unhook papasan's clothing from the barbwire. Then I looked into his face and said, "Hey, it's Ho Chi Minh!" The brothers looked closer at him and laughed and agreed with me. If this wasn't Ho himself, then he had to have been a close relative because he had the same narrow face, deep lines around his eyes, and the exact same style beard and moustache as Uncle Ho. We got him out of the wire and decided we'd better take him up to see the captain because by now it was just about dark, and we couldn't let him continue on the road toward the village. The three of us walked him up the little embankment back into the NDP, and I noticed he looked pretty frightened too, not knowing what we were going to do to him. The four of us came up to the captain's tent, and the orderly called for him and when he came out I said, "Well, Captain, the war's over. We've captured Ho Chi Minh!"

1968. David Sorensen.

The captain laughed and we were dismissed, but I kept my eye on what was going on even after I got back to my platoon's position on the line. The first thing that Headquarters did was to bring a medic over and get him patched up. Then they fed papasan and after, laid out an air mattress for him near the captain's tent. The men on watch at Headquarters were to keep an eye on him throughout the night. The next morning at first light, my buddies and I went down to the road to drag the three rolls of barbwire off it, and while we were there, the Headquarters people escorted papasan to his bicycle. The old man bowed to each of us, got on his bike and went off, pedaling slowly down the red dirt road toward his home in the little village.

Good story, right? Well years later I told this story at a Vet Center in California while I was in a PTSD group therapy session. There were about seven other men in the group, and when I finished one of them said, "We would have just shot him in the wire and left him out all night for the other villagers to see the next morning." More than a few of the other men in the group agreed, and it made me think how lucky I was to have been assigned to the Big Red One because in the 1st Division, from Division Command down to the company level commanders, platoon and squad leaders, they never condoned mistreatment of unarmed civilians. They never asked us to burn someone's hooch down or shoot unarmed people. We never raped anyone nor did any other atrocity that would have brought shame upon ourselves or the division. That just wasn't done and thankfully it was made clear to all of the 1st Division soldiers that, if done, there would be an immediate prosecution. So like I said, I feel blessed that I got to serve in the 1st Division, where people knew right from wrong and where we never lost the sense that we were the good guys, American soldiers.

Walking Point

Kenny Gardellis

November 1967. Sometime in November, I am drafted into walking point. Lima Platoon's infamous point man Johnny was short and went to Di An to process home. I get O'Conner's M-14, ammo, web gear, and compass. I take that job with pride and try to do a walk like Johnny would. I am leading the company of 100 men, and they count on me to get them around safely – or at least for me to find the enemy before they find us. As point man I accept and give into the inevitability that I will or can get shot and die in Vietnam or cheat death and make it out of this war and back to the world in one piece. The life expectancy for a point man is three to five seconds when the first shots are fired. I really don't know what will happen while walking point, but I get peace of mind by giving up the fear and terror of the possibilities of mortal danger. Once the fear is buried and I accept the job as point man, I get a weird peace that allows me to be in the moment and to hear and see more accurately the awareness of my mortality. I feel more at home in Vietnam. I can concentrate on staying alive. It feels like home to me. The downside of that is that I don't write home much; I never feel I am ever going home. I cannot see the light at the end of the tunnel.

Supply Sergeant

Ervin Fox

I served with the 1st Battalion, 18th Infantry, Big Red One, Charlie Company. I went to Vietnam in June 1967 and left June 1968. My duty assignment was that of 76Y40, which is supply. I was responsible for ordering and receiving the supplies for C Company and making sure they got to the field and the men. You can't be good at what you do in the field unless you have the right equipment. I worked closely with the first sergeant and the CO. It takes a lot of men to do the job and do it right. I also worked with the battalion clerks and the division clerks.

Charlie Company was my first priority, nothing was more important than those guys in the field. My CO made sure I didn't forget that. I had two guys who helped me a lot: Bob Duncan and Joe Boland. They helped me find things we needed and were a big factor in getting them to the chopper pad for delivery to the field. Bob Duncan was wounded at Loc Ninh. Then he became Charlie Six driver for a month or so.

Charlie Company, like all the men who fought in Vietnam, needed the best equipment we could get for them, no matter how we got it. Sometimes we had to trade some things that we had for something someone else had, but in the end it worked out great for

Di An Base Camp. 1967. O'Brien left, Ervin Fox right.

both sides.

There were times when spirits were low because Charlie Company was always getting the action. When another battalion or one of the 1/18's companies were hit, we were called in to assist. When our buddies were lost, we all hurt.

October 29-November 5, 1967 was a few days that will always be in my memories. Not that we lost so many men, but the fact that we were surrounded most of the time. As supply sergeant I had to make sure we had all the ammo and grenades we needed. At least the 1st Infantry Division was great at giving us what we needed. It made my job a lot easier. During one mortar attack I was crawling from bunker to bunker making sure the men had plenty of ammo.

At the end of January the Tet Offensive started, and it would turn out to be a bad month for Charlie Company in 1968. It's like the NVA/VC were becoming a real pain in the ass for everybody, but like always we took care of the problem.

The first patrol I was ever on, which I didn't want to go on, but the CO said he needed me out there for some reason, we were ambushed by a few VC, and our medic was KIA. At the time he was

Loc Ninh. Oct 29-Nov 5, 1967. NDP Nov. Platoon's Area.

shot, he was giving aide to a wounded soldier. He didn't even care that he wasn't armed; he just ran out to the friend who was down. That was another low point in my Vietnam tour.

One other accident that happened was a young kid who was waiting to load the chopper for the field and walked too close to the rear end of the chopper and the rear blade took part of his head off. Never saw the kid again and don't know what happened after that. Just a young kid doing his job in the Vietnam War.

The most important thing about my tour of duty in Vietnam was the bond with the men of Charlie Company. Even today they seem like brothers to me. The war was a bad experience for me, in one sense because I had to see so many people killed, including women and kids, which really took a toll on me. The bad times outnumber the good times, but we made a few good times on our own. And I thank God for all of us who made it home, and I will always have memories of the ones who died in a place called Vietnam.

Vietnam Stories

Paul Douglas Goddard Jr.

Decem ber 1967-December 1968. I got to Vietnam during the Christmas truce in 1967. Although I had orders for the 4th Infantry Division, I was reassigned to the 1st Infantry Division. I reported along with a dozen other new lieutenants to division headquarters at Di An. From there I was assigned to 1st Battalion, 18th Infantry. The day I reported for duty was the same day that George Tronsrue took over command of the battalion. This was late December 1967, and with Captain Annan I went to the Charlie Company area that night.

For the first few days I was an extra lieutenant and stayed with the company command post. My first night in the field, we were mortared. The company first sergeant said I had now qualified for my Combat Infantryman's Badge. On our second morning air assault out, I accompanied Lieutenant Tom Pippin with Mike Platoon. The next day Tom became Charlie Company executive officer, and I became the Mike (2nd) Platoon Leader.

One of the first things I noticed was how fortunate I was to have a platoon sergeant like John Kia. Kia was a Korean War veteran, a Hawaiian, and had been in Vietnam for six months. This was a textbook case of what every new platoon leader would hope for, a seasoned veteran to get you started off right.

Within a few days of my taking over Mike Platoon, Captain Phil McClure became Charlie Six. For the next few weeks we did operations out of Lai Khe. On the 15th of January when we had returned to Lai Khe for an overnight stand down, I received my promotion to 1st Lieutenant. January had a lot of patrols, very little action, but it was good to sharpen my skills. The last day of the month Captain

PFC Alan Moniz, Sp4 William Dearing, PFC Terry Hill, Lieutenant Doug Goddard (pointing), PFC Louis Caraballo, on ground: PFC Craig McKinstray, and Sp4 William Hart.

McClure came to us and briefed the platoon leaders that we were on Yellow Alert. He said that we don't know what is going to happen; we just know that something is going to happen tonight. Our company was on the perimeter at Lai Khe, and early in the morning some rockets started coming in. The Quarter Cav (1st Squadron, 4th Cavalry Regiment, or ¼ Cav) moved out early in the morning going to the district capital of Ben Cat. This was the start of the Tet Offensive, January 31, 1968. The week before, our battalion had surrounded Ben Cat one night to allow the village to be searched the next day. I was on the perimeter right where it closed. As we were coming together to close the perimeter, about a dozen VC started running out. We could not fire at them for fear of hurting our own people. They dropped some of the food they were carrying. For the next few days, our company did patrols around Lai Khe.

In early February we took a long helicopter ride to the Saigon

Saigon. January 1968. Tet Offensive.

Water Plant. From there we rode trucks to the Saigon Power Plant located in the Saigon suburb of Thu Duc. (This was a diesel-powered electric generating facility that provided 60 percent of the electricity for Saigon.) For the next 55 days we secured the Saigon Power Plant and did operations in the surrounding area. During 52 of those 55 days we had some enemy contact. Oscar Platoon (mortars) was located on the roof. The three rifle platoons were in sandbag bunkers surrounding the building. At times a platoon was airlifted into an area looking for a North Vietnamese Army regiment. If contact was made, then reinforcements would be airlifted in. The time at Thu Duc proved our competence and commitment to the mission at hand. Every rifle platoon took its turn on company point. Every rifle squad took its turn on platoon point. Mike Platoon became very proficient on the proper way to do an air assault. During our time at Thu Duc, Mike Platoon went on 34 air assaults. Of those 34 air assaults, we were the first platoon on the ground in 19 of them. I was so proud of the way Mike Platoon accepted responsibility and performed our mission. We had our routine down. Everyone knew we depended on each other. I felt like the platoon responded to me not because of the rank on my collar, but because they knew I was competent and looking out for them.

A great example of the competence of Mike Platoon was demonstrated one night on ambush. Captain McClure in midafternoon gave me an ambush location alongside the highway from the Saigon docks to the Long Binh supply base that the supply convoys used at night. We were to face away from the road to guard against an ambush by the NVA. To me, it was just a spot on the map. We were attached to an armored cavalry unit for the night. This unit, commanded by a major, was responsible for the highway security for that night. Tanks and armored personnel carriers picked us up at dark. We were taken to the location, and the armored vehicles made a lot of movement, creating a diversion. We were able to roll off of the vehicles and into the ditch alongside of the road without being noticed. I had arranged the ambush in the normal way of two semicircles forward, one to the rear. Just as everyone got into position, Platoon Sergeant Kia called me on the radio. He was speaking so softly, I had difficulty hearing him.

He said "We are next to an NVA alpha bravo. Thinking I had not correctly heard him. I said, "Say again." When he repeated, I looked back where he was in the rear position but could not see the enemy. I reported our situation to my CO for the night. The major, since we had never worked together before, did not believe me. I asked him for illumination, so we could see them and expose them, and also suggested that he stop the supply convoy before it started until we could clear the enemy ambush. He refused. Later I found out that our own Oscar Platoon was monitoring our radio and was ready to fire illumination. As the first truck in the convoy approached, I watched as a red ball of an RPG went just in front of it. The truck stopped, and I ordered Mike Platoon to fire. I used the one handheld flare that I had to provide a little illumination. Now the major believed me. We tried to pursue the NVA, but this was difficult since there were several houses in the area. He called us back to the road. The convoy proceeded without incident, and we remained in that location for the rest of the night. I always thought this was a great example of our skill – that we could set up an ambush right next to an NVA ambush without them knowing about it.

At the power plant we were able to have our full mess hall setup and have a hot meal for breakfast and supper. One of things I remember the most was an older lady who would come and stand beside the garbage cans where we dumped our trays at the end of the meal. She looked 75 but probably was not yet 45. As each person came toward the garbage can, she would extend her hand for each tray, and with her hand would scrape the leftovers into a cardboard box. This food she gathered was not for animal consumption; it was for humans. The sight and thought of this bothered a lot of the men.

Although the 1st Infantry Division was at full strength, I never had more than 32 men in Mike Platoon. The correct level would have been 44. This has puzzled and frustrated me even to this day. The division was full strength, but the rifle platoons were short.

On another air assault from the power plant, Mike Platoon landed on a paved two-lane road that was unused and not near any civilian population. The platoon moved out to meet our checkpoints. The goal was searching for large NVA units that had come to Saigon for the Tet Offensive. We crossed a stream, and I was concerned about the high bank and no activity, so I had the platoon move down the stream. Shortly, I had a strong feeling that we needed to be out, so I immediately had the point squad climb the bank on the far side. Once everyone was out of the stream and up the bank, we cautiously made our way on the trail on that side. SP4 Richard Granados, a very competent soldier, was walking point. We came upon three NVA. Granados immediately threw a hand grenade, and they fled. One fired back. Immediately Dogface Six, Lieutenant Colonel Tronsrue, was overhead in his helicopter. He brought in a gunship, and I directed their fire. We suspected that we had come across the outpost for a larger unit. Since there were no units available to reinforce us, we were instructed to go back by a different route to our pick up location, and we would be extracted there. Because of our continual enemy contact, Private First Class James Warke had started carrying a whole case of 64 rounds for his thump gun each day. When we arrived at our pickup point, I instructed him to fire at the tree line on the other side of the stream, where we had last seen the three NVA to deter the enemy from following us. On this patrol we had been accompanied by a "Kit Carson scout," a former NVA soldier, who assisted us on our patrols. He was disappointed that we did not pursue the three NVA.

As platoon leader, I was never concerned about the men properly caring for their weapons or carrying the 14 magazines of ammunition and the five grenades required by our Standard Operating Procedures. The men were good soldiers, and Kia was a top-notch platoon sergeant. In addition to the men, I always knew that we had excellent support from higher ups, like when Lieutenant Colonel Tronsrue gave assistance with gunship support, or another day when Mike Platoon was in contact when the assistant division commander, Brigadier General Talbott, later division commander, came on my radio and asked if I needed any fire support – artillery or gunships. General Talbott was overhead and had noticed our contact. Since we were already clearing the area, I assured him that we were okay.

On one of our first company patrols out of Thu Duc, we came to an open field with a tree line on the other side. Following standard procedures, we thumped the tree line on the other side. The VC were spooked and fired back at us. Since we had no indication of their size, Captain McClure had the company pull back, and a heavy artillery mission was called on the tree line. After this, we easily crossed the open area, saw a dead VC, and proceeded the remainder of the day without incident. There was evidence that a large VC unit had been camped nearby. We found several protective holes dug in the side of dikes. That was the first dead body that I saw.

One night at Thu Duc, we were told by battalion intelligence that the power plant would be hit that night. Later in the evening an ARVN officer came by and told Captain McClure that we would be hit at 1:00 a.m. Captain McClure had the whole company up and in our firing positions. Right on time they opened up, firing their recoilless rifle. Fortunately for us, they were firing at an electrical substation located less than a half a mile away. The substation had a lot of power lines around it, while the power plant looked just like another multistory brick building. Immediately Oscar Platoon spotted the flash from the recoilless rifle and fired on them, stopping the attack. The next day Captain McClure and other platoons from Charlie Company swept the area, collected intelligence, and found several dead NVA. The NVA had tortured and abused several villagers in the area.

March 5, 1968 started like any other day at the power plant. The men were happy and getting ready for patrol. SP4 Richard Granados had just returned from R & R and had 60 days left in country. Charlie Company was going on a company reconnaissance in force patrol that day. We left the power plant with my platoon in point. After some distance we came to a large clearing of about 30 acres. I took the right column and crossed the opening. We moved toward a small building in the center. Platoon Sergeant Kia took the left column and followed along the tree line. Today was the largest number of men I had in Mike Platoon at 35, which included a squad of ARVN Marines. SP4 Richard Granados and SP4 Bill Sullivan shared their experience and competence with a new man they were training on point. About halfway through the open area, shots rang out on the tree line. The ARVN Platoon leader behind me was immediately shot through the heart. Platoon Sergeant Kia was at the point of contact. The company medic came forward and confirmed the death. Leaving my RTO, SP4 Gene Anderson, I moved across the field to the point of contact. Kia cautioned me about crossing the stream where the fire had originated. I fired my AR-15 as I crossed the stream. Upon reaching the point, several men were watching medic Brian Montgomery as he treated SP4 Richard Granados, who had been severely wounded. I contacted Charlie Six and asked for a medevac. Although an ARVN medevac had evacuated the ARVN lieutenant, it took a very long time for our medevac to arrive. Dogface Six was overhead at the time to assist. By then it was evident that Richard had been killed. As his spirit left, I prayerfully thought about the pain his family would experience soon. We later found out that we were operating at the edge of a map sheet and the helicopter pilots had been given the adjoining map sheet, which put them some distance away from us. Although I was disgusted and frustrated and disappointed, Dogface Six, Lieutenant Colonel Tronsrue, insisted on the medevac coming in. Upon our return to the power plant, I immediately inquired about Granados. He had been killed instantly. A sad day for Charlie Company, Mike Platoon, and me personally. This was the only person I lost in my year in Vietnam under my command. (When I returned home, I sent a copy of the photo I had taken of Richard that morning to his mother.)

Even though Charlie Company was on the other side of the world from Washington, D.C., we were very aware of what was going on politically during this time at the power plant following the Tet Offensive. I had learned that President Johnson had ordered a bombing halt to encourage peace talks. This was disturbing to me because ceasing the bombing gave the enemy an opportunity to receive new equipment in South Vietnam with which to fight us. We suspected that the North Vietnamese were able to get even more advanced weapons, as we were seeing evidence of that in recent weeks.

In June I found evidence of some of this new equipment recently possessed by the VC. Charlie Company was located at

Thunder Two. At this NDP there was one bunker in the center of the front gate. At night a gate would be opened on either side of that bunker, and armored vehicles would go up and down Highway 13 looking for any VC/NVA putting mines in the road or setting up an ambush. It was scary to be in that center bunker when the vehicles would go racing out and back into the NDP during the night. The tank and APC tracks would come just inches from the bunker. On this night after the vehicles had returned and everything was quiet again, SP4 George Felker observed movement in the darkness outside ahead of that center bunker. I instructed him to fire the thump gun at that movement. After that, all was quiet. The next morning my platoon did a sweep outside the gate and found equipment on the ground. While some of the items were old and worn, like a small shovel, a grenade pouch, and a canvas sack full of rice, there were items that were brand new: a chest magazine carrier for an AK-47, a backpack containing two RPG rounds, and an infrared scope. The latter was made in Czechoslovakia. The items had been cut off the person who carried them, and this person probably was dragged away to avoid our detection. I had hoped to bring the infrared scope home, but someone in intelligence kept it after I had turned these items in to be catalogued.

Charlie Company did conduct an amphibious operation while in Thu Duc. We were taken by truck to a South Vietnamese Navy (SVN) dock in the Saigon River. We loaded onto SVN World War II Higgins boats to move to a very large island in the Saigon River called VC Island. Our mission was to land prepared for an assault but also to take a medical team to a village on the island. When the ramps on the boat dropped, we stormed to the shore without a shot being fired. After securing the village, our medical team set up on the grounds of an old destroyed cathedral. Throughout the day people came forward for medical assistance. An eighty-year-old woman was carried on a litter by four village elders. We departed VC Island

April 1968. VC Island. Giving medical help to the people of the island. Captain Phillip McClure with hands on his hips and 1SG Maynard Ward far right with back to camera.

feeling good about our day there.

Leaving Thu Duc, we went on a battalion reconnaissance in force mission. Mike Platoon was first on the ground, followed by the remainder of Charlie Company. We were assigned to be at the tail of the battalion column moving out of the landing zone. As the rest of Charlie Company came in, I saw one helicopter crash and break the tail boom off. At the time I did not know who was in the helicopter. I saw a squad leader from another platoon walk away from the crash. Later I found out it was our company command group. As the battalion moved out of the landing zone, Dogface Six called me on my radio, which was something very unusual. He indicated that Charlie Six was evacuated after the crash, and I was now Charlie Six. I moved up the company column to find the command group and took over the company until Captain McClure returned to duty days later. Twice, Captain McClure came to our NDP to discuss with me the activities of the day before returning to the rear at night.

Later in April our battalion was attached to the 11th Armored Cavalry Regiment commanded by Colonel George S. Patton, the son of the famous WWII general. Charlie Company was split up with each rifle platoon going with an armored company. We were operating in an area that we were told no non-Vietnamese had been since the French in the 1950s. We found a road network that had been made through the jungle for the NVA to move supplies toward Saigon for the Tet Offensive. Tree branches had been tied together to obscure the road from aerial observation. Operating with the armor was scary for us because we were infantrymen. The armor made a lot of noise; we liked being quiet and operating with stealth. I was fortunate that the armor company commander accepted my recommendation that Mike Platoon sweep the ground away from the armored vehicles. It was during this operation that Lieutenant Edward Mello and SP4 Kim Deter were killed while with another armored company. Later that day Mike Platoon returned to the site where these casualties had occurred. We were instructed to sweep the area to see if any NVA remained. This was the most scared I had been during my Vietnam tour – taking my men in an area where a few hours before two of our men had been killed. The ground was black from fire, but fortunately for us the enemy was gone.

The battalion reassembled the next day and airlifted to a site on the Dong Nai River. We were to secure a pontoon bridge that was being constructed to allow the 11th ACR to cross upon. At this time Lieutenant Smart had taken over Charlie Company as Captain McClure had become Battalion S-3 Operations Officer upon the death of Major Albert Maroscher. We had some tanks in our perimeter and several vehicles with the engineer bridge company, making us an attractive target. Intelligence had intercepted an enemy message saying to destroy our site at all costs. The bridge was built, but the armored unit had not crossed yet.

That night when we heard the thump of mortars firing in the distance, everyone sprang into action. We had prepared our bunkers and fields of fire well. We were very seasoned by now. Immediately I called in our listening posts and left my bunker to make sure they were all in their position and no longer out in front of us. I ran into SP4 Edward Pilieri coming in from LP, and we both fell down. When he and I returned to my command bunker, all of Mike Platoon was in firing position. The tank located next to my bunker started firing flechettes from their main gun. Earlier in the day I had had the tank move back a few feet so that our firing port was clear in front. Unfortunately, the flames from his main gun were coming into our firing port now. I stood on the step at the back of the bunker and yelled from the top of my lungs for him to pull up, that the noise and flame from the main gun was getting us. While he was moving the tank, a large piece of shrapnel from a mortar round hit me flat in the nose. My verbal response was so strong that Platoon Sergeant Kia grabbed my ankles and pulled me down into the bunker. Fortunately, it was just a bloody nose. I had also received some shrapnel pellets in my back. Several of the vehicles belonging to the engineers were destroyed, including a fuel tanker. The battalion perimeter held, and the attack was repulsed. Several wounded had been taken to the aid station near the battalion CP. The battalion surgeon, the chaplain, and several medics were attending to the wounded. I spoke with SP4 Jeff Clark from our platoon, who had a broken arm. He only had a few days left in country; I said goodbye to him. When I returned to the Mike Platoon area, I was told to report to Lieutenant Colonel Tronsrue at battalion CP.

When I reported, Lieutenant Colonel Tronsrue said that Lieutenant Smart had been wounded and would be evacuated in the morning. He said that I would be Charlie Six in the morning. I visited Lieutenant Smart at the company CP and discussed the change. He had a 3-4 inch wound on the left side of his chest that was no longer bleeding but clearly would have to be stitched. The next morning I took over Charlie Company as we cleared the area and moved out to a checkpoint where we were met by trucks to be moved to a location on Highway 13, north of Lai Khe.

Thunder Road was Highway 13. Several numbered, permanent defensive perimeters, called thunders, had been established along the highway. Companies would rotate to these perimeters regularly to provide security by establishing outposts and checking for mines along their sector of the highway. At night, armored units from the perimeters periodically would run the roads. From 9:00 a.m. to 5:00 p.m. each day, truck convoys would use the roads.

While on Thunder Road, Captain McClure visited Charlie Company to counsel me about the command. He wanted me to know that I was in command of the perimeter, even though we were securing an artillery battery that was under the command of a captain. I was in charge, because I was the infantry commander. Captain McClure also met with the artillery commander to make sure he also understood the chain of command and that he would be protected by Charlie Company. We also had a platoon of five tanks attached to Charlie Company for security on Highway 13 and within our perimeter at night. Charlie Company remained on Thunder Road until Sam Downing joined the company as Charlie Six, and I returned to being the Mike Platoon leader.

Previously at this location there had been a large battle when the NVA had attempted to overrun this NDP. The Howitzers had been fired point blank into the wire. Daily the company went on road security along Highway 13. After the convoys, Charlie Company returned to the NDP each afternoon. At night we would send out an occasional squad-sized ambush.

In preparation for my tour of duty in Vietnam, I wanted to make sure that I was very competent in the use of the map and compass. In all my training I excelled in these activities and felt very confident in the field. While company commander on Highway 13, one night I saw mortars being fired at division headquarters at Lai Khe. I got on top of one of the bunkers, sighted the flashers from the enemy mortar tubes, called in the azimuth from my location to the mortars

Outside of Lai Khe. 1967. NDP.

location as well as the estimated distance. In less than a minute, fire was coming from the Lai Khe base falling directly on these mortars. The artillery knew where I was, and they knew where they were. My information allowed them to triangulate and fire upon the enemy mortars. I remember thinking that this was really cool and worked just the way I had been trained.

In June of 1968, Lieutenant Colonel Gillis succeeded Lieutenant Colonel George Tronsure as Dogface Six. Upon my return from R & R, Lieutenant Colonel Gillis summoned me to his CP. By now I was a senior first lieutenant with considerable field experience in our battalion. Lieutenant Colonel Gillis told me that he wanted to make me a roving company commander. Now that Captain Sam Downing was at Charlie Company, I could go to Bravo Company, where he expected a captain to be available for that assignment in two to three weeks. Then I would move to Alpha Company to be company commander there until their replacement captain became available. I responded that I did not think this was fair to the men or to me, to move a person around every few weeks from assignment to assignment. One could not develop the necessary relationships to command. His response was, "Huh, are you just afraid of getting shot.?" I said, "I have survived now for seven months. Let's see how you do!" Looking back, I can't believe I was that cocky.

In June, Major General Keith L Ware, our division commander, sent out a request for recommendations for a new aide-de-camp. Lieutenant Colonel Gillis nominated me from our battalion. This was an exciting opportunity to have clean sheets for the rest of my tour, although I did have reservations about leaving my men. Without an interview, I became the brigade commander's nominee a few days later.

The next week I was scheduled for an interview. We were in the field off Highway 1. I was going to the rear to clean up, where a new uniform had been prepared for me. The helicopter came in, and just as I put my foot on the skid to get in it, Company 1st Sergeant Maynard Ward yelled out, "Lieutenant Goddard, wait! Don't go!" As I stepped back, the helicopter lifted off without me. 1st Sergeant Ward said that General Ware had already made a decision. The son of a good friend, a new lieutenant from the States, had been selected to be his aide-de-camp. I was very disappointed to not even get an interview. At the time, Charlie Company and the whole battalion were in a very muddy, wet location along Highway 1. There would be no clean sheets in my future. Later, General Ware and his complete staff were overseeing an operation from the air when his helicopter was shot down September 13, killing everyone on board. Many times I have thought of this to remind me there was something else God wanted me to do. As I have grown older, I have thought about 1st Lieutenant Steven L. Beck, who got the job as aide-de-camp, and about his family.

Sometime in August, Lieutenant Colonel Gillis assigned me to be Headquarters Company Commander. With concern, I left Charlie Company without a replacement for me as Mike Six. Staff Sergeant Campbell remained as platoon sergeant, and a platoon sergeant was moved from another platoon to be platoon leader.

My last three months or so in Vietnam were spent in Di An in this assignment as HHC Commander. One of my responsibilities was to be the paymaster for all the officers in the battalion. I went monthly to each officer to give him his pay. Sometimes I went by helicopter and also by jeep, with my driver, Sergeant Carter, on payday covering the complete length of Highway 13. I was responsible for all battalion personnel in the rear, including 30 Vietnamese women who served as mess hall workers. The women kitchen police (KP) were paid a dollar a day, which made them some of the best-paid workers in Vietnam. A few times members of Mike Platoon came by to see me when they were in the rear and told me they missed me and appreciated me. That meant a lot to me.

Over the years I have kept up with a few people that I served with and have reconnected with more in the past three years at the Charlie Company reunions. I continue to search for the men who served in my platoon. Even though I may never see some of them again, the bond we formed in combat is still strong. I have fond memories of the men I served with in Vietnam. They were good soldiers doing what their country asked them to do.

Village Seal

David Gilbert

On January 28, 1968, C Company went on a village seal in the vicinity of Ben Cui. This village we thought had a lot of VC/NVA activity in it at night. Lima Platoon went one way and Mike and November went the other. We were going to meet up on the back side of the village. These kinds of patrols were really dan-

Outside of Saigon. 1968. Kenny Rucker heading into a village.

gerous, especially at night. I guess the VC were going in at night to see their girlfriends and family.

Tommy Mercer was on point for Lima Platoon, and I was the fifth man back in the patrol. This night I was carrying the M-79 grenade launcher. I never had to walk point and had no interest in doing so. It was more action than I needed.

Mercer heard noise about 30 feet in front of him and went into his kill mode. Being on point, he only had a second or two to decide what he was going to do. He opened up and killed two VC right off. O'Be started firing with Tom. I started firing and so did Kenny Gardellis. Kenny carried an M-14 and also was a point man. Mercer was saying later, anytime you have VC 30 feet in front of you, there is no choice, you had better shoot.

We were worried about the other part of C Company that we were to meet up with, hoping we had not shot in their direction. O'Be said he knew exactly where we were and also where the rest of the company was. After the firing was over, the guys up front went to check out the area. At this time I heard Mercer yell, "O'Be, I've got an M-16 over here." He was in shock. He thought he had killed an American. Then he was relieved when saw it was a VC. We had killed four and wounded a few on that little firefight, but anytime you are getting shot at is dangerous. I was doing all I could do to fire as many rounds as I could. Of course, with the rest of C Company to our right, I had to fire my M-79 where the VC might be running to.

We finally met up with the rest of C Company and settled down for the night. We still had to be awake and watching out for the VC in case they came back and tried to get some revenge of their own. My main concern was a mortar attack.

Once again C Company did what we had to do, and it worked out great. I was glad when that village seal, or ambush, was over. Even the simple ones are bad. A few of the guys received medals for that night, but like the guys said, it's not about the medals.

The Beginnings of the Tet Offensive – January 1968

Colonel George M. Tronsrue Jr.

Dogface Battalion (well, most of it) was asleep. As one of the main maneuver units of the 1st Infantry Division, we had just finished a demanding series of what were usually described as "security patrolling missions" in South Vietnam. That jargon term doesn't begin to describe the individual soldier's hectic daily life. He was regularly and, often, unexpectedly, in harm's way as we all tried to prevent the Vietnamese communist soldiers from taking control of the Republic of South Vietnam by force.

Our battalion, the 1st Battalion, 18th Infantry – nicknamed "Dogface" for its current radio call sign – was at about full strength (750 soldiers, more or less). We were assembled in the vicinity of the division headquarters at Lai Khe, about 50 or 60 miles north of Saigon. We had just completed several days of security patrolling in the jungles nearby. We had been moved to an inner, secured area of unoccupied buildings for a short rest. It was after midnight on January 31, 1968, and all was quiet where we were. Our soldiers were enjoying a peaceful sleep, under actual roofs, warm and dry – an altogether rare happening. Our battalion TOC, the nerve center of the battalion, was peacefully calm and relatively still – again, a rare happening.

All of that changed in an instant. Our command net radio operators, routinely listening to the traffic between other, distant operators just to keep abreast of what was going on around us, began to hear very unusual and increasingly excited talk. Clearly, something out of the ordinary was happening in the areas far to our south near Saigon. As they sat up and listened more closely, it became clear that the radio traffic was apparently describing heavy enemy shelling and even enemy ground attacks in the Saigon area. "Tet 68" was in full swing, and life for Dogface Battalion would never be the same, again.

Shortly after news of the attacks in the Saigon area reached the division headquarters, the battalion was given a unique role. We were ordered to assemble at the Division Forward airfield at Lai Khe before daybreak. Within minutes we were airlifted by helicopters to the division rear area at Di An, on the northeast edge of the Saigon area. After landing, we paused for a few more minutes and then flew on to Tan Son Nhut Airport, the main airport for Saigon, and the Army headquarters location.

Our leading helicopter, which carried the battalion command group (the commander, command sergeant major, and two radio operators) was met by a frantic junior U.S. Army staff officer who was, as we understood him, from the Army headquarters. He gave us very abbreviated, even cryptic, instructions: "They're over there!"

and we promptly moved in double-column attack formation out the southeast gate of the airport. There was no one else around. Before we had moved more than about 200 meters into the built-up area outside the gate, we were fired upon, and the next hours were busy ones.

Charlie Company 1/18th Infantry

Phillip E. McClure

When I assumed command of Charlie Company in January 1968, I was starting my second tour after spending nine months as an advisor to a Vietnamese recon company and three months with 2nd Squadron, 5th Cav Air Mobile. I did not know what to expect, except that I was assigned to a unit with an outstanding reputation. Within days, I was operating in a rubber plantation observing the outgoing commander. What struck me was the discipline of movement and the reaction when light contact was initiated. This unit had good leadership and had been battle tested. Over the next few weeks my initial observation proved to be correct. We were constantly on patrol making enemy contact almost every day. Most were company-sized operations. However, a company-sized operation was normally three platoons minus; rarely did the number of soldiers exceed 70-80 men unless augmented by the weapons platoon. I always placed my command group behind the lead platoon to allow for better command and control and to give the trail platoons maneuver room. I was questioned about this as many times we were in the middle of the firefight along with the lead platoon. However, in tight jungle terrain, you have to see and feel the fight to affect supporting fires. We had a number of hard rules: don't walk on

1968. Kim Deeter, Captain Phil McClure (retired as a colonel), and Bob Quimby.

dikes, don't cross open areas without sweeping the sides and testing forward, never push forward if the lead point sensed trouble until the Arty FO was ready with supporting fire, then move slowly, and finally, we would never leave anyone behind.

Things emphasized in my initial briefing: obey orders, accomplish the mission, wear a steel helmet, shave every day when possible, and take care of your soldiers. I added to that: bring everyone home. The key was not to do anything stupid, reduce risk, and communicate, up and down. We had exceptional leadership down into the squads, and most important, everyone looked out for each other. The confidence we had in each other and in the way we operated greatly reduced risk.

The air tactical scheme was make contact, call in the world, and then sweep. We did not try to slug it out one-on-one without first punishing the enemy with supporting fires. With few exceptions, we dug in with overhead cover at night, ceased all noise activities prior to dark and listened for the sound of incoming mortar rounds. As a result, we did not fear mortar attacks nor sustain serious injuries from them. On the lighter side, I remember being called to the bat-

April 1968. Heading to VC Island with the Vietnamese Navy.

talion CP while having leg cramps. I was moving a short distant, very slowly, when I heard the mortar rounds coming out of the tubes. I can assure you that one can run really fast with a leg cramp when properly motivated. On returning to my bunker, everyone was inside arguing about who was supposed to bring the cake and coffee (left from evening chow) into the hole. To get into the bunker, I had to bring these items with me.

We had the attitude that we were all in this together and together is how we would survive and get out. The stress of going out the wire nearly every day was hard on everyone. No one resented the effort to dig in, and everyone helped dig. The confidence in our preparations to remain overnight allowed everyone to relax just enough to take the intense pressure off, yet not so much to lose alertness. We learned from past experiences: good and bad. Shortly after assuming command of the company, I was told about a helicopter crash that resulted in serious injuries from unsecured equipment. We talked about what to do in case of a future helicopter crash. A few weeks later my command group went down, but nobody was seriously injured. The pilot did a terrific job getting us down through the jungle canopy, everyone piled on the equipment, and we all walked away. Needless to say, the helicopter was a total loss. At all levels we tried to talk about what happened and how to do it better next time. But to be effective, everyone had to have a chance to participate and learn. No one was worried about hurt feelings; we just wanted to improve and do better next time.

We were attached out and assigned to occupy Thunder 3, an NDP on the main supply route, when the Tet Offensive started. The battalion was deployed to Saigon to help retake Tan Son Nut Airbase and secure MACV headquarters. Subsequently, the battalion was sent to secure the Saigon Water Plant. We (C Company) rejoined them there, but in turn were sent to secure the Saigon Power Plant. As was often the case, we were apart from but under the control of our battalion. We were not too happy about this until we found out they had flush toilets and unlimited hot showers. After months in the field without, this was heaven.

We operated out of the power plant, patrolling to the north and east of Saigon in one of the staging areas for the Tet Offensive. Out the wire daily we were operating in rice paddies and nipa palm-lined canals with many small hamlets in what we called the Factory Area. We were not allowed into the factory compounds even though we constantly made contact in and around these facilities. Most of these contacts were with North Vietnamese and VC elements that had survived the Tet Offensive. Although the daily patrol routes were sent by coded radio messages, I usually changed them based on my knowledge of the area, not repeating the same routes of ingress and egress, and what we experienced from previous fights. The battalion commander, Lieutenant Colonel George Tronsrue, had the confidence in us to allow me the latitude to make these changes. Throughout this constant daily action, everyone stayed focused on the job, allowing us to recover numerous weapons, documents, and prisoners.

Our soldiers never lost their concern for the innocent casualties of war. In one case, we set up a night ambush in an area where we usually made contact. Early on, either a Claymore was detonated or an RPG was fired at us, but nothing else. I decided to stay in position. About an hour before daylight, the rear security element engaged targets. It turned out that two young kids had violated curfew and rode their bikes into the ambush. Doc did his best, but to save the kids we needed to get them to a doctor; our guys felt horrible. It was still dark, and we did not know if the enemy was still in the area, so we could not risk a helicopter evacuation. The battalion commander launched an A/Cav element to evacuate the kids; they survived. Certainly it was a risk. However, the impact of not helping them would have been everlasting on my soldiers. Shortly after this, the company conducted an operation with the Vietnamese Navy to sweep an island in the river, called VC Island, and then conduct a medcap for what was left of the population. There was no contact. The battalion surgeon and his medics had a full day, but it was fun and satisfying to work with the people and actually play with the kids.

After the power plant assignment, the battalion moved north to Lai Khe, Division HQs, as they were being rocketed daily with 122mm rockets. A battalion NDP was established northeast of Lai Khe to find and destroy the rocket sites. Operating from this base, C Company conducted a RIF and made contact that midafternoon. We ambushed an enemy party coming down a trail and recovered five or six weapons and documents. Although I was in command, I had an observer: the new battalion operations officer out on his first field operation. After the initial contact, he wanted to push forward, but I made the decision to pull back. We had been moving through an area with many booby traps and fresh signs of enemy activity. I did not want to get caught in a big firefight with darkness approaching and in an area where we could not maneuver because of the booby traps. A couple of days later, as C Co led a battalion movement, we again made contact just as the enemy was firing rockets at Lai Khe. It was a base camp area set up for approximately 500 troops. A small force was in the camp, providing security for those firing the rockets at Lai Khe. They were in spider holes and trees and managed to fire one rocket over us during the firefight. We recovered a number of unfired rockets, weapons, and several food caches.

From Lai Khe we moved north again up near Song Be. We were looking for what was referred to as the Adams Road, an enemy main supply route first reported by Special Forces recon units. We located it and were amazed at the road and bridges that had been built under the jungle canopy. We also uncovered signs that tracked vehicles had recently used the road. The enemy was reported to have Soviet PT76 light tanks, and we now had proof. Although we had an armored cavalry troop with us, they had no HE rounds and that made for a very uncomfortable night. For this operation the battalion was attached to the 11th ACR for a push through War Zone D to the Dong Nai River to establish a bridgehead to receive and cross the main ACR force. Our operations officer was killed here, and I became the battalion operations officer. Battalion conducted a short recon patrol, with our attached A/Cav troop and C Co's Lima Platoon (1st platoon), in the direction that we would be moving toward the river.

Lieutenant Emmett Smart, C Company XO, came forward to take command of the company. I tried to monitor the company's activities, but it was not necessary as Lieutenant Smart was more than up to the task. Unfortunately, he was wounded in the Dong Nai River fight. We had a bridge company attached along with the Cav troop to establish and secure the Dong Nai River crossing. As I recall, we had most of the tactical bridging equipment in country. The crossing was established with our B Company securing the bridgehead on the west side of the river, and the Battalion(-) with attachments on the east side, with the bridging trucks inside the perimeter. The 11th ACR was late getting to the river, so the crossing was completed in the late afternoon. As soon as they crossed, the engineers began breaking down the bridge, leaving B Company on the west side of the river as the task force prepared for movement eastward. The plan was to bring B Company across the river by boats to join the battalion the next morning. With all the trucks and armor in the perimeter, we made a fat target. That night, they hit us from the north, mainly in Charlie Company's area, with mortars, RPGs, and small-arms fire, igniting numerous trucks and at least one APC. Company C held but sustained a number of wounded, including Lieutenant Smart. The company did a great job that night. Soon after, Captain Sam Downing assumed command of Charlie Company and continued their tradition of excellence.

Anyone who says he was not scared or fearful has not been in combat. The record shows these men experienced this every day, faced whatever was asked of them, and went on taking care of each other. They accomplished daily feats that most can only imagine or read about, as daily actions. Each and every one of them overcame their fears and pushed on. They are all heroes, recognized or not.

Power Plant

Tom Mercer

After a long stay on Highway 13 we finally got a break, we thought. We were now headed to Saigon, to the power plant. Our commanders expected the NVA to take it out. That power plant carried 85% of Saigon's power.

When we landed, we were all in shock. The bunkers were on concrete, with no mud around anywhere. We thought we were in heaven. Lima's part of the plant was in front of the building. We had

Power plant. M-60 machine gun on bunker.

real tall bunkers, and they were built really well. In the bunker we had a 50 cal. machine gun and a bazooka on the top.

The good part didn't last too long because we had ambush patrol that very night we arrived. These ambushes were a lot different than the ones out in the jungle. We headed into the outside of the city and set up in a vacant house, which I thought was really strange, but the colonel knew best.

Daylight patrols were just about as bad as the ambushes at night. There were a lot of NVA/VC around the area. The NVA were helping the VC fight around Saigon. They had better weapons and better training; so the both of them together made it even harder for us to fight.

Charlie Company were jungle fighters, and this was all new to us. We had to adapt fast. Patrols were really scary because we didn't know where the enemy was going to be – in a house, behind a wall, in a truck, we just didn't know.

There were a lot of rice patties all over the place, and that was always a bad place for an ambush to happen. The NVA didn't care; all they wanted to do was kill Americans whenever they got the

Power Plant. February 1968. Lima's side.

chance. The bad thing for them was when they got close enough to do that, we would kick their ass. They would hit and run. That's what they were really good at.

After every patrol we could get a hot shower in a certain area of the plant. It was like being back home and clean for a few hours each day. The officers made sure we didn't get too comfortable. They kept us busy every day.

Out of a Rambo Movie

Tom Mercer

February 1968 – Power plant near Saigon. C Company was on patrol, and Lima Platoon had the point for the day. Throughout the day we had been taking a lot of sniper fire, and it was starting to tick everybody off. We would walk awhile, and then we would have some more shots coming in, just harassing us. The small ambushes were a pain also. They would hit and run. Now we were mad and ready to kick some ass.

Every now and then we would see clothing that the enemy had taken off and thrown away along with some of their supplies they didn't need. The last piece of clothing we found was the pajamas; then all they had on was their underwear. They were trying to get as light as possible so they could run faster. We stayed right on their heels and were getting closer. I remember Kenny Gardellis and I talked about the clothes we were finding along the trail. We couldn't figure out the reason why. We found out later.

Captain Phillip McClure was being real careful not to follow the VC into any kind of ambush. He had our platoon leader call in the big guns to soften them up. After that we would move ahead, thinking we were going to have a chance to stop them. I know after every small contact, there were always blood trails leading off down the trail; so we knew we were getting to them. The big guns were leaving lots of signs of wounded VC that had been dragged away.

After chasing them for a few hours, we came up to a mound about eight-feet high. The point man stopped the patrol and said to get down. He was looking right at a VC who was looking right back at him. The VC was standing in water up to his knees with his AK-47 above his head. The VC was lucky the point man didn't have a hair trigger like some of us old timers. We radioed the platoon leader the situation and asked what to do. The lieutenant said not to shoot the VC. Captain McClure was on his way up to the point, and we needed information from him. We were trying to figure out how to shoot him without getting in trouble.

We were thinking that just one VC had stopped and given up, because he was yelling, "Chieu hoi," which means he surrenders. The VC were in real bad shape. They looked like they hadn't eaten in a week, at least not a lot.

The interpreter came up to us, yelling orders at the VC. Finally he put his AK-47 down in the water. We went and picked it up. We made the VC strip down to his underwear and squat down in the water with his hands still on his head. The VC started pointing to his right and saying something to the interpreter. The interpreter raised his rifle in that direction. Then we got a big surprise. Four more VC stood up out of the water with their AK-47s above their heads. They were under the water, breathing through a piece of bamboo about eight inches long. It was like something right out of a Rambo movie. Scared the hell out of all of us. We had never come across anything like that. These VC were really lucky. I believe if we had messed with that one VC, the other four would have shot all of us. They must have just been tired of the war and fighting. For them it was over. I also believe they were glad to get out of the water because it was full of leeches, which hurt. Lots of times, when a VC or NVA was captured, he became an interpreter for the American troops, and most worked out just fine.

Every time I see an Army movie or a Rambo movie, I think of that patrol. We are lucky to be alive.

Kids at the Power Plant

Tom Mercer

At the power plant, we noticed every day that two kids always came around while their mamas were selling things to earn money. Joe Boland named them Batman and Robin. We thought the kids were so cute and innocent. Well, we found out they were cute but not so innocent. They would steal anything they could get their hands on. We always kept an eye on them so they couldn't get away with anything of value. We all got quite fond of the little guys.

This one little kid was about four years old and had a friend maybe a year younger and would always go for the candy, but what they didn't know is that we had laid it out for them to take. We were on to them. Sometimes we would put C-Rations with the candy to help feed the family. Each night we could see them coming, and we would turn our heads so they would think we weren't looking. Those kids took everything we had out for them and would take off running.

The next day here they come like nothing had happened, and we would ask them, "Did you guys steal our candy and C-Rations?" The one kid looked at us and of course said, "No. Me no steal from GI." Then he would point at his friend. 'He steal your candy." Then he would walk a few feet and turn around and yell, "We steal GI's candy." Of course we took off after them, acting like we were really trying to catch each one.

The kids were so proud that they stole our candy and laughing so hard with tears rolling down their cheeks because we couldn't catch up with either one of them – although sometimes we would catch one and throw him from one guy to another. Believe me, those kids loved it. The day we left the power plant, they were really sad, even though we told them that more GIs would take our place. That seemed to pep them up a little. I always wondered what those two kids must have really thought about us after we left. Joe Boland and I have talked about the little guys a lot in the past few years. I hope they both made it all right when the North took over Saigon. The 1/18th loved kids from all over Vietnam, and we were not baby killers as some people seemed to believe. Most of us still lived at home when we joined the Army. We were used to playing with our younger brothers and sisters, so this made us feel at home. After we left the power plant, we all missed Batman and Robin.

The Acquisition

Alan Moniz

I recall one evening when we were in Lai Khe, our platoon leader put together a small squad of guys to go over to the airstrip where a line of deuce-and-a-halves were parked, ready to go out on a convoy. The purpose of this mission was to acquire some M-14s, which were in short supply and carried on the trucks by the drivers. We were to remove an M-14 and replace it with an M-16 and several clips of ammo so as to not leave anyone without a weapon to defend themselves.

We had to be very careful not to be seen or caught during this operation, as there could be serious consequences for making such

Alan Moniz

an acquisition. We were very successful – we picked up five M-14s, which we desperately needed to increase our firepower on patrol. Once we had them, we had the supply sergeant get us some operating rods, handles and springs to make the weapons fully automatic. They are normally semiautomatic weapons.

I carried this memory with me for over 40 years and wondered if I had imagined. Then at one of our reunions, a speaker brought up the acquisition of the five M-14s at Lai Khe. At a later meeting one of our officers mentioned that there were five M-14s unaccounted for, and they wondered where they had come from.

This was just one of the kinds of things that took place in Vietnam. When supplies ran short, you had to adjust and do whatever was needed to resupply your unit and complete your mission.

The Men of Mortar Platoon

Tom Mercer

There were some men who served in Charlie Company who did their job well but never received much credit. These were the men in Oscar Platoon, our mortar platoon. They were the backbone of a rifle company. We could never have accomplished what we did without them. Mortar platoon was always there with us, patrol or ambush, and always had our backs.

Many people don't know what goes on in a mortar platoon. Although they were behind the rifle platoons, they were definitely in the line of fire. They went on LP/OP. Many times they were the primary target. If the NVA or VC could knock out the mortar tubes, it would be a lot easier for them to get into our NDP.

When the NVA or VC were firing their mortar rounds, our mortar platoon was doing the same thing. They would be up and running around while the rounds were coming in. I often looked back, and I couldn't believe what I was seeing. I told everyone that those were either brave or crazy men, but they did it in every firefight we were in.

My friend Michael Shapiro was in mortar platoon. Michael said the thing they worried about most was bringing down fire too close to our men and killing someone on ambush or patrol and even

December 6, 1967. Men of Mortar Platoon, 1968, Toacum, Luevano, Robert May.

at the NDP. Michael would always come around after a firefight to make sure his friends were all right and that none of his rounds hit too close for comfort. I once had dirt blown all over me from a mortar round that got a little close, but I thought of it as a security blanket. If I was getting dirt on me, then the enemy was getting the rounds dropped right in their laps where they needed to be.

Mortar platoon didn't go on many ambushes, but they were always "there" from a distance. They knew exactly where we were at and could drop a round anywhere we needed it. I can vouch for that because I am living to write this story. Oscar platoon did go on LP and OP.

Ambush in April 1968

Tom Mercer

This was a big ambush that started off a bad month of April 1968 in the Lai Khe area right after the Tet Offensive ended. We were hoping the NVA/VC had headed back up North where they belonged.

We left the NDP right before dusk. There was just enough daylight to see where we were going. By the time we got to the ambush position, it was dark. Lima Platoon of C Company had the ambush. Our platoon sergeant was in charge of the ambush this night. He was new in country, and we weren't sure about him yet. He was about 45 years old – maybe even older. I thought he was way too old to be in the field and in the jungles of Vietnam, but he was there, and there was nothing we could do about it.

We went out about ¾ of a mile from the NDP; we had 12 guys with us. The rest of the platoon stayed back to cover our positions and man the bunkers. At the time we had only about eight men per squad. Once we got to our ambush site, we checked it just to make sure there was nobody there with us. The full moon helped.

It was a great spot for our ambush. The road split off to the right and to the left and joined back together on the other side about 20 yards. In the middle of the split were some trees and a lot of bushes, and that is where we set up. This night we took one M-60, four M-14s, five M-16s, and two M-79s; so we had plenty of firepower. The platoon sergeant wasn't too sure of how things were to be done on an ambush; so Kenny, Smitty, and I helped him with the setup and placement of the Claymore mines and trip flares. We added trip flares because sometimes the NVA/VC would sneak up and turn the Claymores to face us.

The LRRP unit of the 1/18 was in the area and had reported a lot of movement where we were set up. When the LRRPs said there were NVA/VC in the area, you had damn well better believe them.

After we had been on the ambush for a couple hours, Lieutenant Smart radioed and said the LRRPs would be heading our way and not to open fire on them. They were trying to get the enemy to engage in a shootout and reveal their position.

We were set up. The guys had their detonators right beside them ready for firing. Kenny and I explained to the guys how we would blow the Claymores if we had to fight our way out. We did not want to fire our weapons, mainly because that would give our positions away. A small area like our's could be fatal.

Kenny Gardellis was a squad leader while he was a private first class. Taking charge as a PFC was not an unusual thing in Lima Platoon in April 1968.We had very few sergeants in our platoon as we started that month off.

Most of the night had gone by with no action. It was about 2:30 a.m., and I saw a lot of flashes down the road. Of course I was thinking, "What the hell is that?" The LRRPs were walking down the middle of the road, tossing white phosphorous grenades to each side, hoping the enemy would be there. As the LRRPs came up to our ambush site, they acted like we were in it even there. They just nodded and kept on moving down the road. They went a short distance and then went into the woods and were gone. The enemy were nowhere close to us. They were by the NDP getting ready for an attack.

Suddenly we heard that sound nobody likes to hear: thump – thump – thump. The NDP was getting mortared, and the VC were dropping them down the tubes fast. Then the ground attack started. Lieutenant Emmett Smart radioed the platoon sergeant and told him what was going on. He also informed him that the enemy was

1968. Choppers coming in to pick up C Company for an eagle flight to another NDP or an Ambush? 1SG Maynard Ward in foreground.

heading our way.

I guess Lieutenant Smart didn't think the platoon sergeant knew what he was doing. He had been unsure; so the lieutenant relieved him from being in charge of the ambush. He didn't wait because it might have gotten someone killed. Now Lieutenant Smart wanted to talk to me. The platoon sergeant was new in country and this was his first big ambush, and he just wasn't sure. Lieutenant Smart told me what was going on back at the NDP. I was told to get ready for some action; the enemy was on its way toward us. I then told the guys what Lieutenant Smart told me and to get ready.

It was now around 3:15 a.m., and the guys were hearing a lot of noise. The noise got louder and louder. We were as still as you could be on an ambush. Not a sound was made. Even our breathing was slower and silent so not to be heard, not taking any chances. I looked around and could tell the new guys were getting real nervous about being there, and so was I. The noise was real loud now, and for a while we thought they were going to walk right into our ambush position.

Finally they stopped and were either setting up in an ambush themselves or taking a break. I think it was an ambush since they had just hit our NDP. I guess they thought we would be sending a patrol to follow them, but that's not the way we worked. Their ambush was about 30 yards from our position, and I'm sure they had a good spot. We were one up on them because we knew where they were, and they were not aware of us.

Both ambushes were within 30 yards of each other, and that was way too close for comfort. All it would have taken was for someone to make just a little noise and the fight would be on. And as anybody who was in Vietnam knows: a fight at night is no fun at all.

Around 4:30 a.m. the guys had been up all night, and I was worried about people falling asleep, including myself. The enemy must have grown tired of waiting and was getting ready to move out. We still didn't know how many there were and really didn't want to find out. But I couldn't believe what I saw next – they started moving out to the road. They were moving to the side of us, and we could see a few of them. Gardellis had his men set up to cover the side where the shooting started.

One of them tripped on David Gilbert's Claymore wire. David grabbed it and blew his Claymore. Then another man blew his Claymore, which was the right thing to do. They were right up on top of their position. Then all hell broke loose. Four more Claymores detonated, and it sounded like a big war was going on. Now the M-60 gunner fired a lot of rounds into their position. The thump gunner was blasting away. A couple more Claymores were set off, and everybody was firing, trying to make us sound like a platoon of men on our ambush. We had four Claymores left, to be used when and if we had to leave, or if we needed them now. We were getting a lot of return AK-47 rounds fire, but nobody was getting hit. They must have been firing on the run. Thank God there were no RPGs fired at us.

Everything was over about 4:45, and the enemy was gone.

Those 15 minutes seemed like forever. Lieutenant Emmett Smart was on the radio with me and let me know he had a platoon of men ready to go out and get us if we needed them. Unlike the LRRPs, we couldn't get out as fast, because we had so many men. We stayed up and alert for a couple more hours, and then it was daylight. Finally Lieutenant Smart said he was sending the rest of the guys out to get us and check out the area.

Once we saw the platoon, we got out of our position really carefully, just in case the enemy was waiting for us. We checked out the area and found four dead enemy and blood trails where they had dragged off the wounded or dead. The area where David Gilbert detonated his Claymore had two dead VC. In the wood line was another dead VC, and one more about 20 feet from our position. He was torn all to hell with M-16 and M-14 holes. The two that David Gilbert killed with his Claymore looked like they had been through a meat grinder. When we were finished, we had a big surprise for us. One of the men killed turned out to be a Russian advisor. Well at least he was dead, and nobody would have to worry about him again.

That was one scary night but the guys did great, and the way we had the weapons divided up worked out well. We had a lot of new guys, but that night they were broken in the right way. Some of them were in a state of shock, but they were alive. Most of the guys on the ambush had been in country for a while, but a few were FNGs.

When we headed back in everybody welcomed us with yells and handshakes, and the company CO gave us a job well done. (The platoon sergeant got a good ass chewing.)

Lima Platoon killed four NVA/VC on ambush without losing anybody. Firepower and a great setup of the ambush position is what got us through that night.

Rocket Day – April 10, 1968

Tom Mercer

E very time we left an old NDP, there were thoughts running through everyone's mind: "Where the hell are we going now and what are we looking for?" Of course, we would always hear bits and pieces of information, not knowing what was true and what was not. This would drive us crazy because we were always thinking the worst was coming.

This time we heard we were going searching for rockets 12 miles northwest of Lai Khe in an area called the Rocket Belt The VC had been firing rockets at our division headquarters every day and every night. We had heard General Ware was really getting pissed and wanted it stopped. This area had lots of NVA roaming around and big base camps set up for the NVA and VC.

We got word down through the chain of command that our battalion commander, Colonel Tronsrue, was going with us on this one. This meant we had to do things exactly by the book. Colonel Tronsrue wanted us to be careful; there was a lot of NVA/VC in the area. That's always good to hear right before you head out on a patrol. The colonel was a no-nonsense type of commander, and he took his job seriously. He didn't like getting anybody hurt or killed. Captain Phillip McClure was in charge of Charlie Company. He was a tough and rugged type of commander and did not mind getting right in the middle of the fight with his men. He would have been a great point man.

Charlie Company had point: Lima Platoon up front, 1st Squad was in the lead, and I was point man. Kenny Gardellis was the second man in the patrol and right behind him was Rusty Little and Kimble Myrick. Rick Morrow was new in country. He was in 2nd

April 10, 1968. Tom Mercer with 122mm Russian rocket, Rocket Day, 12 miles outside of Lai Khe.

Squad and didn't know what to expect if we got into a firefight. If it happened, this would be his first. At the time I was an acting squad leader who was walking point. Kenny and Rusty behind me were two good men to have up front with you. Throughout the day we changed point men to rest each other. Kenny was a well-used point man. Throughout the year he and I walked point quite a lot for Lima Platoon. Kenny knew how to keep his cool when a firefight started.

I didn't think it would be too bad because Colonel George Tronsrue was going with us. Why I would think that, who knows. We had walked for about two hours doing our regular maneuvers. We would walk a while, then stop and send out some cloverleaf patrols, which was a patrol outside a patrol. The smaller patrols sometimes would help find ambushes that were set up by the VC and waiting for the main patrol to walk into. They were like a safety precaution. Some people had the impression that the NVA/VC were stupid; they were not. They had been fighting for years, and they were great at setting up a good ambush.

This day we had "Fritz," a scout dog, with us and his handler from the 35th Scout Dog Platoon. Fritz was a black German Sheppard and he was a good one! The handler would have to get the dog used to the point man, but we weren't allowed to touch the dog. The dog gave us a feeling of security because if there was a VC in the area, Fritz would sniff him out. We found this to be true later that day.

I was hoping for a nice easy day, and not seeing any NVA/VC at all would have been nice. Sometimes we had a feeling that something bad was going to happen, which is the way I felt that day. With everything that goes through your mind, you can't even imagine what may happen on a patrol in a free fire zone. As point man on this particular day, I had my eyes and my ears working as hard as possible. I stopped the patrol immediately for anything that didn't look right. The longer I was on point, the more nervous I was. I was considered one of the better point men in Charlie Company, and we had some really good ones at the time: Kenny Gardellis, Bill Sullivan, Robert Norris, Lon (Smitty) Smith, Thomas Cone, Rick Rossi, and Tom Murphy. We all did things the right way.

After another hour of walking we got word that rockets were being fired in Lai Khe toward the 1st Division Headquarters. That was one thing we didn't want to hear. That means we were getting close to what we were looking for and didn't want to find. Things changed and everyone was in a serious mood. I noticed everybody was looking around like this might be the big day for all of us. Will this be the day we walk into a big ambush like so many soldiers had done in Vietnam over the years?

Our platoon leader at the time (KIA) was new in the country and had never seen combat. He kept telling me to be careful and keep my eyes open, reassuring me that everything would be okay. I had a hard time understanding how he could tell me everything would be okay. I had been in country for nine months and had been

1968. Tom Murphy – one of Mike Platoon's great point men.

in a few firefights and battles. He had never been in a firefight. I knew he meant well, but mainly I was hoping he could keep it together if we had contact.

Captain McClure was about 17 men back in the formation. He liked being up front where the action was. He had all the guys' respect.

After we got the news of the rockets being fired toward Lai Khe, we started moving again. We had walked for about one hour, and Fritz, the scout dog, started acting up. There was definitely something in the area, and he was detecting it. When the dog handler told me to keep my eyes open and be really careful, I knew something was going to happen.

Again we started moving and maybe went about 40 yards when Fritz acted up again. This time the dog handler said we have to stop and check out the area. I passed on this information to the platoon leader. The lieutenant said we were not stopping and to keep going. I knew we should not keep going, but we did go a little further until we came to a clearing. Again I stopped the patrol and told the lieutenant we have got to check this out before we go across the open area. The lieutenant still did not want to do a cloverleaf; he wanted to keep going. I stopped and refused to lead the patrol across the clearing without first checking it out. I was still on point, and Kenny Gardellis was behind me and then Rusty Little and the dog handler, Sergeant Miguel Soldana. The lieutenant was a little ticked off at me, but I was right.

The platoon sergeant spotted two rockets set up and ready to be fired. Against my judgment, I took a few more steps. The VC let Kenny, Rusty and me go into the open area with the dog handler and then began firing at us. We were pinned down and couldn't move at all. Rusty Little was shot in the hand and fell to the ground. Kenny Gardellis was trying to make sure Rusty was okay. Even being shot in the hand, Rusty Little was still doing what he could to kill the VC that were shooting at them The dog handler laid his body on top of Fritz so he would not get shot. The dog was going crazy at this point. The VC would love to have killed Fritz; they hated our scout dogs.

This is when I decided to get up and run toward the VC, firing my M-14. The VC must have thought I was crazy for doing so. My action took me forward toward a log and closer to their position. I was trying to get behind cover. If I had dropped to the ground when the firing started, most likely I would have been shot and killed. My instinct took over, and I really don't remember everything that went on after that point. I didn't take a lot of time to think about what move I was going to make; it came naturally. My buddies were down and needed help. Any one of the guys would have done the same thing I did, but I was the one who was there.

I felt a little relieved when Kenny Gardellis had come up where I was and helped me get out of my bad position. We were firing our M-14s and throwing hand grenades toward the enemy's position. Kenny and I were doing all we could to make sure the VC didn't shoot the guys that were still pinned down, including the dog and the handler. I'm glad Kenny was able to help me because at the time all I had for protection was a log and my M-14.

There was at least a squad of NVA/VC, even though we only killed one. There were signs of blood from either wounded or dead VC that were being dragged into the woods. One VC could not have carried those big rockets through the woods by himself, so we know there was at least a squad.

I remember when the firing started, there were men from the other platoons coming up to help. That's what made Charlie Company so great. Lieutenant Doug Goddard, who was Mike Platoon leader, was one of the first ones up there. I remember Joe Boland from supply trying to put out the fire in the grass that started when the first rocket was shot off. We had to get the two rockets that were still there out of the fire. Thomas Cone and I grabbed the rockets and got them away as fast as we could. We were hoping they would not go off in our arms. It's a good thing Cone had already cut the wires away from the rockets.

The rockets were 122mm, Russian made. They were six feet long and would definitely go a long way. They were set up on a real unusual platform consisting of two wooden sticks about one inch in diameter and three feet long. They were bound together with wire, six inches from the end forming a makeshift tripod. The power source was eight flashlight batteries that were wired together. They had an American Claymore detonator that was attached to the batteries to fire the rockets. The 1st Infantry Division intelligence spokesman said this was the first time we had found the VC employing such a launching platform. The VC used this type of platform because it was a fast set up, and they didn't have to carry a heavy tripod through the jungle when they moved to different sites.

I guess the things we did that day helped keep a lot of men and ourselves from getting wounded or killed. All the on job training that we had up to this point really worked out. I know General Ware was glad that we got the rockets. Then maybe Lai Khe could sleep better at night.

We got word that General Ware was coming out to the field to check out the rockets that had been haunting him for months. When he got to the field, he was wearing two .45 revolvers and had a white German Sheppard that stayed right by his side. I think Fritz the scout dog was liking the general's white German Sheppard. True love in a war zone. When they informed me the general wanted to talk to me, I was scared to death. The general told me what a good job we had done and promoted me from a private first class to a sergeant E-5. Before I got the promotion I was an acting squad leader as a PFC. Nobody was aware both Kenny Gardellis and I and a few others were going to receive medals for our actions.

Colonel Tronsrue and Captain McClure were proud of Charlie Company that day, and I'm sure they got some pats on their backs from General Ware. One thing for sure, it was pats on the backs that were well deserved for the battalion commander and Charlie Company commander.

April 11, 1968. The next morning was going fine until we headed for the chow line. We heard the worst sound you could ever hear in a war zone: thump . . . thump . . . thump. Men were yelling, "Incoming," and in just a few seconds mortar rounds started hitting all around us. The VC had us zeroed in. We hadn't had time to dig our bunkers like we normally did the night before. One round hit the chow line, wounding a few men and killing our machine gunner who died instantly. The mortar rounds continued coming in for five minutes, maybe longer. A lot of men were cut off from our bunkers and could not get to them. So we had to find cover wherever we could. I remember hugging a big tree and hoping for the best. The mortars were covering an area of about 30 yards left and right.

After the mortar rounds quit coming in, we found out how many men were wounded or killed. Everyone was in a panic mode. We still had to keep our guard up, watching out for the VC that may be trying to attack us from the tree line. This is when the squad leaders started doing their jobs getting their men back together in case of a ground attack from the wood line. We put out LPs all in the wood line, more than usual. The NVA/VC blew their chance when they didn't hit us right after the mortar attack.

Once we got the situation under control, we called the medevac choppers in and got the wounded airlifted out. Each platoon sent out a small patrol around the area looking for any VC stragglers. We wanted to make contact now to get revenge for the men we had just lost in the mortar attack, and we had a lot wounded.

Once the patrol got back in, we went about our business as usual. A few hours later we caved the bunkers in and were getting ready to get airlifted out and head to Lai Khe. We found out we were going to have a few days rest and an awards ceremony. The few days' rest was the highlight of it all, but I was glad that Charlie Company was recognized for something we had done as a team. I felt kind of weird receiving a medal for something that was just part of my job. The real heroes to me were the men that were killed that day and the ones wounded.

We then headed back out to the field to a place called Song Be. The last time we were there, a lot of firefights took place.

My First Time Getting Shot At

Rick Morrow

I remember the first time I had rounds go over my head and real close to my body. I was scared to death. Being new in country I really didn't know what to expect, but I found out. It was April 10, 1968, 12 miles outside of Lai Khe in an area called the Rocket Belt. We were on patrol when I heard shots cracking over my

1968. Rick Morrow at Fire Base.

head. I jumped behind a tree, but to my surprise everybody was running right past me. I was thinking, "Are these guys crazy?" I decided I had better follow them, so I wouldn't get stuck back there by myself. We ran all the way to the front where the firefight was going on. Someone told me to toss a grenade into the tree line where the VC were at, and I did. The only thing is, I didn't throw it hard enough, and it didn't go far enough. My first big mistake. I got a lot of dirty looks from all the old timers, and they weren't too happy. This would be the day I would see my first dead VC, and it didn't break my heart. After I settled down, it bothered me a little, but what could I do?

The next morning our machine gunner told me to go and get some chow. I wasn't very hungry, but another guy and I went to eat anyway. After we got back from the chow line, the machine gunner went. As we started to eat, we heard that awful sound of mortar rounds coming in: thump – thump – thump. We all dropped our plates and headed for the bunkers. The first rounds hit in the middle of the chow line, killing our gunner and wounding a lot of the guys. Then the VC started walking the rounds up and down our bunker line.

This is when it really hit home for me. I became a man that day along with a lot of the guys. I saw my first dead VC and my first friend killed. That really opened my eyes to what was to come in the next few months in Vietnam.

Rocket Day

Kenny Gardellis

April 10th, 1968. Charlie Company is sent on this special mission to find the VC who are firing Russian-made 122 mm rockets into Lai Khe. Our company, four platoons to include Oscar Platoon, our mortar platoon, is flown out to a clearing a few miles outside Lai Khe. Tommy Mercer and I are walking point. Rusty Little was up front with us, also a scout dog named Fritz and his handler, a guy named Miguel Serano from NYC. He speaks with Mercer and me before the patrol and tells us about his dog and how the dog will react and respond to the scent or detection of VC. It is a hot day, dry, and the point element stops suddenly. We are getting signs that the dog is sniffing the air; so we stop to listen. We are walking very slowly and quietly. Things are tense and very hot. We have been on the patrol for about five hours total when Fritz alerts us. We all hit the ground just as AK-47 bullets come flying over our heads. We fire back and forth, and there is a lot of gunfire. There are about six VC firing at us. Serano does his job – protecting his dog – and lets us grunts do what we do best: kill the bad guys. Mercer and I run forward to try and see where the VC gunfire is coming from. We are firing our weapons and moving forward, hiding behind trees and fallen logs.

Meanwhile, Mercer and I keep the bullets flying and dodge the AK-47 rounds. I hear a bullet come past my ear while I am lying behind a log. That's a sound I won't forget: a bullet with my name on it that misses me by merely an inch or two. Before the company can catch up to us through the thick vines and branches, Mercer and I can tell where the gunfire is coming from. Mercer comes over to the log I'm firing from and whispers to me that we should get closer to the enemy. So he tells me to start firing my M-14 on automatic in the direction of the enemy so he can run closer and hide behind another log. I fire as he maneuvers. When he gets to his log, he waves for me to move up to him. He fires his weapon at the VC as I run toward him. As we make our way closer, we kill one VC, but another one fires a rocket off just 20 yards from me. It is awesome and surprises me. It takes off in slow motion, a six-and-one-half foot rocket with a three-foot bright orange flame bursting out the end.

That particular rocket hits the hospital in Lai Khe, but no one is killed. At the rocket launch site, one VC lies dead and the others run off. The rest of our company finally catches up with us and searches the area, eventually finding a rocket or two with batteries wired in series used to ignite the rocket fuel. The batteries are US Army batteries, and the tripods are made of tree branches. The captain calls in some choppers to pick up the rockets, and the rocket fire into Lai Khe stops for a long time but not forever. Lai Khe earns the name "Rocket City."

That night we dig bunkers and set up an NDP. We have left the dead enemy lying where he fell. We send an ambush out and an LP. As soon as it gets dark, about four enemy soldiers come back to retrieve the body of the dead one. The LP guys see them and fire their weapons, which eventually both jam. I am monitoring the radio when I hear the LP ask the captain if they can come in from the LP since their M-16s jammed. The captain says, "No!" After further discussion the guys on LP say, "We're coming in, sir," and they blow their Claymores and run back to our NDP. From then on we are all awake most of the night expecting trouble. It is quiet until about 5 o'clock in the morning, when the sound of enemy mortars leaving the tube breaks the silence. I hear someone yell, "Incoming!" I am in the chow line with a paper plate and plastic knife and fork when I hear the mortar rounds leaving the tubes: thump, thump, thump. I run and hide behind a giant termite mound as the mortars start landing inside our perimeter and all around me. It is pitch black, and I can't see anything except the flash of explosions on the ground. Our mortar platoon comes to life and starts returning fire.

It only lasts minutes, and only about 10 to 12 mortar rounds drop inside our perimeter, but the damage is done: my friend and machine gunner, Thompson, is killed. Several more of my friends are wounded as well. It is very scary and sad. I'm glad I didn't get hit or killed, but we sure miss Thompson. He was a good, strong, and reliable machine gunner who kept the gun clean and was not afraid to use it. The damn enemy was not far from us and probably launched its mortars from the same place they had their rockets that we had captured the day before. Thompson is found on the ground with his helmet on. Apparently the mortar had landed right next to his head and blew holes through the steel helmet and into his skull, killing him immediately. The general flies in that morning with a big, white dog by his side, and we notice his uniform starched with razor sharp creases and how clean his boots are. He speaks to us and is sorry to hear about our casualties. He asks to see the point men who captured the rockets, dismantled the rocket launch site, and killed a VC. Mercer and I go over to talk to him, and he congratulates us on finding the rockets.

The next day Charlie Company humps out of that death trap area and eventually gets to Lai Khe. After a day or two, Mercer and I walk into a small museum on base and see one of the big old rock-

ets we found a week earlier. There is a plaque that says it was captured by Charlie Company 1st of the 18th – that's us! While Mercer
and I are just looking around, we hear a clerk talking to someone
about how he was there when the rockets were found. Mercer and
I stand by the clerk and listen for a moment. Finally, Mercer gets
pissed and jumps in this guy's face yelling at him, "You're full of s**t!
We nearly got killed capturing that rocket! You better shut your face
and get your story straight." There is no way that we fought the
enemy for that rocket only to have those REMFs take credit for it!
We tell that clerk the real story, tell him not to forget it, and storm
out of there.

The next day the general pins a Silver Star on Mercer, and a
Bronze Star with a V for valor on me and Rusty Little for our actions
against the VC and their rocket launch site. Sometime after the
ceremony General Ware says to our captain and the whole company something to the effect of: "You men are so good. We knew you
were the ones that could find that rocket site, and you did an outstanding job. Because you guys are so damn good, we've got another operation for you to go on tomorrow. Intelligence reports say the
VC and NVA are in this area near Cambodia, and we want you guys
to team up with the armored cavalry on their tanks and armored personnel carriers and go find the enemy and wipe them out." We
were really hoping for a break after finding and capturing the rocket. An R & R at the beach, a steak dinner, or even just a chance to get
a real night's sleep is more what we are hoping for. We certainly
aren't up for another hairy operation in the boonies. However, early
the next morning we saddle up and fly out of Lai Khe to meet up
with an armored Cav unit, perhaps the 11th Cav.

Thomas Cone

Tom Mercer

There was and always will be only one Thomas Cone. If he saw someone a little down, he would come up with something that would cheer up the person. Tom came up with more crap than I ever heard before, and we never knew what to believe. But at least there was one person who did a good job in keeping us in good spirits.

There actually was serious side to Thomas Cone. I remember April 10, 1968. We call it Rocket Day. We were 12 miles outside of Lai Khe, a place called the Rocket Belt. We were looking for NVA/VC positions that were firing rockets into Lai Khe at the 1st Infantry Division Headquarters.

The patrol was moving right along, and everything was going great until we came to the clearing and our platoon sergeant spotted two rockets that were ready to be fired. One rocket went off, and later we got word that it had hit Lai Khe. I was on point that day. We stopped the patrol and that's when the firefight started. We ended up with one man (Rusty Little) getting hit in the hand with an AK-47 round.

When the firing had stopped, Cone and I went over to the rockets. Cone cut the wire on one, and I grabbed the other. The grass under the rockets had caught fire, and we had to get it out. He was on the ball that day. I remember him telling all the guys that we are going to get hit tonight or tomorrow. He was right. We got it the next morning.

The next morning, April 11, 1968, as we were in line getting breakfast that they had brought out to the field to us, we heard the mortars being firing and coming in. The attack didn't last too long,

1968. Thomas Cone, Robert Norris, Lauren Coleman, Alan Moniz – sitting down, waiting to airlift to NDP.

but it was costly one. The VC had our position zeroed in perfectly. We had one man killed and several men wounded.

Thomas Cone was one of the wounded, but that didn't slow him down. Instead of staying in his hole, where he should have been, he moved around helping the other guys. He put a few men on the medevac chopper so they could be air lifted out. Thomas was not telling jokes during those few minutes. In fact, I had never seen Cone move so fast in the months we had been there. That's the kind of guy Thomas Cone was. He joked a lot, but when he was needed, he was there. A good man to have around.

Ambush in Rice Patty

Kenny Gardellis

Later in April 1968, I remember being on an ambush with about eight guys one night. We were lying in a dry rice paddy, pulling guard duty. We all had our Claymore mines out on a paddy dike, just four feet above us. Someone touched me and whispered, "Shh, VC." I grabbed the detonator of my Claymore. I heard footsteps and the conversation of the enemy as they walked into our kill zone. Six came out of the village walking on top of the rice paddy dike. One of our guys whispers, "Get ready to blow your Claymores." Then the lead Claymore blew with a tremendous BOOM, and we all clicked our detonators at once. We're all lying on our stomachs, just listening. Word passes down by whisper, "We got 'em." We decide to wait until dawn to see the carnage. In the morning our squad leader says, "Let's see what we got." There is no lack of curious volunteers. I am not one of them. I figured after all the Claymore blasts, there would be little left to identify. Six bodies are found and six AK-47s.

I was glad those VC didn't see us or our Claymores on the trail. They were out that night looking for us, but we spoiled their plan. I feel sorrow and loss for friends of mine that were killed in earlier battles with the VC, possibly even this group we stopped tonight. Revenge is a powerful thing if used right. Can you imagine the emotions of loss, sorrow, and emptiness when a good friend whom you share a bunker with is killed – gone forever? All of the guys in Lima Platoon are very tight. We share stories of our lives in "the world." When a friend is killed, it leaves a hole in the heart. It doesn't bleed out until many years later. We dig bunkers as a team, we live together, wake each other up for guard duty, smoke a cigarette together, eat C-rats, share water, share stories of home-towns, families, mothers' cooking. We see photos, get letters, share care packages of food and other goodies. We carry the weight of their memories with us. God bless our fallen comrades from Vietnam.

Red Ants Psychological Casualties

Kenny Gardellis

With the exception of Loc Ninh, I like walking through rubber tree plantations because I can see and it is shady. There are irrigation ditches that the enemy could use as a place to conceal themselves – then ambush us. But the ditches provide cover against hostile enemy fire if they miss you the first time. Of course, my best days are when I don't have to walk point. Everything about not walking point is like a day off. The point man of another platoon breaks the bush and hacks a trail through thick jungle vines with thorns and the red ants that make their nests in the leaves of trees not too high off the ground. I remember those ants well. They use saliva to weld two leaves together to make a nest, and if you break through the bush and knock a nest down, these ants will attack you fiercely, bite, and hold onto your flesh with their pincers. You have to stop the patrol and start stripping your gear off and your shirt, and other guys have to come over and help you pick and brush them off. They are tough little bastards, and, man, do they sting! It has happened to me more than once, and it's tough because you have to stop and deal with them immediately. Also, usually you're making some noise like "ouch" or "damn" or "son of a bitch" or something like that. Luckily the enemy never hears me during these times; if they do they probably laugh at me and then run off so as not to get shot.

As the months drag on, the heat gets hotter and the red ants get meaner and the leeches have sucked my blood too often. The "f°°k you" lizard is out at night in the dry season – always giving us his or her opinion about what he or she thinks of every GI in the country of Vietnam: "F°°k you." It is worse than Hanoi Hannah.

Psychological casualties are heavy in this war. At one time I am walking point in some fruit grove with Vietnamese family hooches around when I see four VC with rifles. So I fire on them before they see me. I duck behind a tree and reload my M-14. I fired 20 rounds at those suckers but none fall. Instead they run, and we don't see them again. Soon after, as I go back to talk to the captain, I pass the RTO for the platoon leader. I look at him, sweat streaming down my face like at a sauna, and I see he is crying. We look at each other, and he says, "I just can't take it anymore." It was a rough patrol, I'll admit: three or four days of patrols and ambushes, some VC snipers, just s°°t happening every day and night. Nobody got any sleep. So I understand the man's anguish, but I also know he is toasted, his brain fried, his nerves shot. They call in a dust-off to take him back to Di An. I never hear from him or any news about him. He is gone but not forgotten.

Mechanized Unit Patrol

Tom Mercer

We went back to Lai Khe for two days for an awards ceremony and rest after Rocket Day, April 10th. That day took a lot out of all of us. We were not prepared for it, especially losing the guys we did. We had one man KIA and nine men wounded.

After two days' rest we headed out to a new NDP, where we

were attached to the 11th Armored Calvary. Our new location would be around Song Be with lots of NVA around. After a couple of days we went on a patrol with the 11th Armored Calvary. We had a squad on each APC. This patrol would turn out to be a day from hell for the Lima Platoon, Charlie Company, and the 11th Armored Cav. At first we thought it was nice to be riding instead of walking.

Our platoon leader was Lieutenant Mellow, fresh out of Officer Candidate School. He had only been in the country for about two months. Lieutenant Mellow was a good guy and wanted to do the right thing. Rocket Day was his first big action, and he did okay. Being a new officer he depended on some of the old timers to help him out with different things, which we did. The lieutenant wasn't the kind of person who thought he knew everything; he was willing to learn. He said to me, "That's not what I was expecting."

As we headed out on patrol, we were on top of the APCs and not inside. To me that was the worst place to be. I couldn't stand the thought of being inside the APC when a round hits and being hurt or killed without having a chance to fight. Lima Platoon had point so we were up front of the patrol on top of the APCs. Kenny Gardellis, David Gilbert, Tommy Stranno, and I were on the third APC back. Lieutenant Mellow, Kim Deeter, Rick Morrow, and Lauren Coleman were on the fourth APC back. There were a few others, but I can't recall their names.

We had gone on our way for a few miles and came to a curve in the road. The lead tank headed straight into the woods when all hell broke loose. The lead tank was hit by an RPG, which killed the driver and the gunner, and wounded another man inside. The APC behind the tank was also hit with an RPG round and disabled. A couple of men were wounded from that hit. The driver of the APC was hit and died April 25, 1968. After the first shots and rockets were fired, the APCs started tearing up the bamboo on the one side and the jungle on the other. It's a good thing nobody got in the way.

We could not move forward or backward and were taking .50 caliber and AK-47 rounds all over the APCs. It sounded like a swarm of bees flying all around. I wanted my squad to get off and on the ground with me, but everybody was jumping off on their own. I'm not sure if the captain wanted us to get off or not, but we could have cared less what the captain had to say. No use having any of us killed on the APCs. At the time we were getting off, we were still under fire.

On our right side was the jungle, on the left side was thick bamboo 4-8 inches around, and the .50 caliber was tearing the bamboo apart. The first two APCs were still getting hit with RPG rounds and out of action.

Once we got on the ground, we were still in a bad spot. My purpose for getting off the APCs was to get away from them because that's what the VC were trying to tear up. I wanted my guys to move to the bamboo side of the road, where the VC were. There was a ditch on the bamboo side of the road, and we needed to get there. We would be lower than the VC line of fire and definitely would have a better chance. The guys must have thought I was crazy, but nobody had a lot of time to make the right or wrong decision. Good choices came from remembering what you did in past firefights; bad choices came from not remembering anything.

The APCs were still getting hit badly with RPGs, but at least the guys were safe for the time being. Even though we couldn't stand up yet, we were putting out all the firepower we could. The bamboo was being torn all to hell. The good thing was that nobody freaked out. We had been together for a while and that helped a lot. We knew that none of us squad leaders were going to do something stupid. It kind of reminded me of Loc Ninh: lots of firepower from both sides and a lot of men going down. The guys kept on fighting and never thought about quitting.

Once we got in place in the ditch, I moved David Ballard and a few guys down to the right, toward the tank that got blown up. Lieutenant Mellow had sent Rick Morrow and a couple more men to help the guys in the tank and the APCs.

David Ballard always carried more ammo than he was supposed to because he never wanted to run out. With 23 magazines, he wouldn't have to worry about that at all. The limit he was supposed to have was nine.

Once David and the other guys were in place, David, a hunter from Alabama, spotted where the rockets and AK-47s were being fired. He opened up and put 40 rounds into the area. No more rockets were fired after that. Now all we were worried about was the .50 calibers and the AK-47s. David said at one time an RPG hit the APC, the lift door fell down, and four guys ran out bleeding like hell.

Luckily the bamboo was thick enough that the VC didn't have a good field of fire to get at us. If they had set up at the other side of the APCs they would have torn us up. We would not have had any cover at all and no good chance for survival.

Lieutenant Mellow told me he hated to give me an order like he had just received from the captain commanding the cavalry troop. He was told to take his men into the bamboo straight in front of us. Kenny Gardellis and I tried to talk him out of it, but he insisted. Gardellis, Smitty, and I were thinking what a stupid order for the captain to give. Even the new platoon sergeant told Lieutenant Mellow this was not a good spot to enter the bamboo. I had been in the country for 10 months and was not going to do something that stupid and get my men or myself killed. We suggested to Lieutenant Mellow that he call Captain McClure (Charlie Six), but he said this is what the troop commander had ordered him to do, and he had to obey. This was the type of officer Lieutenant Mellow was. He was just trying to do what he was told to do. He had not been in country long enough to know you could refuse an order as long as you had a real good reason. Getting bad orders, like the one Lieutenant

Tommy Strano (left), unknown, unknown.

Mellow just got, should have been refused. I really don't think the troop commander knew what he was getting us into. When men die, officers have to answer to the battalion commander what happened and what went wrong. Lieutenant Mellow was just doing what he had been taught to do: obey orders. If he had been in country a little longer, things might have turned out differently.

Kenny Gardellis noticed a new guy, Tommy Strano, was heading toward Lieutenant Mellow. Gardellis grabbed him, pulled him back, and told him he had better come with us because that's not going to be a good place to be. I took my men to the bamboo area 10 yards to the left of Lieutenant Mellow's position. Deeter, Smitty, and a few other men were with the lieutenant. Going the way I went would put us on the left side of where the .50 caliber was firing. I thought that maybe, if we could come in on the side of them, we would have an advantage, and they would not have a clear shot at any of us. Normally, we would have pulled back and called Captain McClure, he would have called Colonel Tronsrue, and they would have brought the world down on the VC with fire support. Too much was going on, and we couldn't move the APCs.

May 1968. Lauren Coleman, Doc (Medic) Nelson, Lt. Bryon Dinnison.

Once we got about five yards into the bamboo, all hell broke loose with .50 caliber and AK-47s going off. I saw Lieutenant Mellow running into the bamboo to help Deeter, his RTO. The guys were yelling, "He's down. Lieutenant Mellow is down." Lieutenant Mellow was hit helping Deeter, who had just been shot himself. What Lieutenant Mellow did was way beyond the call of duty. One of his men was down, and he went to get him. Not thinking or caring about the outcome of his actions – the act of a true hero. I still don't know why he wasn't decorated.

Rick Morrow was 50 feet to the right of where the action was. He saw what was happening and headed toward where Deeter was. On his way there, Morrow got hit and was down but not out. After the medic fixed him up, he headed back out to the fighting but fell down from the wound to his leg so he started reloading magazines for the other guys.

My guys and I crawled back to where Lieutenant Mellow's position was. The medic yelled for me to go out with him to get the lieutenant. The medic was a good friend, and I guess that's why he called for me to help him. It took four of us to get the lieutenant out of the bamboo because we couldn't stand up at all. The medic, Lon (Smitty) Smith, Ken Gardellis, and I went out to bring the lieutenant back into the road. When you see a buddy down, all of your common sense goes away. All you can think about is helping that person. Any one of the guys would have gone out there to help, but they were busy with their own problems at the time. Strano and some of the other guys were covering us to try and keep the VC down so they couldn't fire in our direction.

We had to stay low going out to get Lieutenant Mellow, and I couldn't even take my weapon. I emptied everything out of my pockets. I knew if I needed a weapon, I could use the lieutenant's. When Doc Nelson and I got to Lieutenant Mellow, he was barely alive. The medic did a tracheotomy – he cut a small hole in the lieutenant's throat and placed a small rubber tube in the incision. This kept him alive for about three minutes as we were pulling him out. His last words were: "Please help me. Don't let me die," but it was too late. He died as we were taking him out. I have thought about that day for 42 years. I will take to my grave those words of Lieutenant Mellow.

David Deeter was hit but made it out and away from the incoming rounds. The medic fixed him up, but Deeter, for some strange reason, ran out to help Lieutenant Mellow. Deeter got as far as the ditch and was shot again, but this time it was fatal. An act of a true hero. Smitty was in the ditch also but could not get to Deeter because of enemy fire. This has bothered Smitty for 42 years and brings tears when we talk about it.

Sometimes it's hard to get a guy to raise up high enough to fire his weapon, but not Deeter. I'm sure all he was thinking about was getting to Lieutenant Mellow because he knew he had been shot also. I am proud to tell everybody about David Deeter. Doing what he did may have saved a lot of lives. You can only do so much when the rounds are coming at you the way they were that day. Everybody was doing all they could do, but Charlie Company always did that. What was so sad is that both of our guys, Lieutenant Mellow and Kim Deeter, died trying to help the other one make it out safely.

David Ballard has told me that after the firefight, the captain of the APCs wanted us to get back and leave the area. David said I told the captain, "No way," and that captain told me I was in a lot of trouble, but at the time I could have cared less. Our platoon sergeant told me not to worry about it. He would tell Captain McClure what happened. As far as I know, he did because I never got into any trouble over it. I don't remember the sergeant's name. I do remember he was really torn up over losing Lieutenant Mellow and Deeter. He was new in country, and I believe this was his first big firefight.

We wondered why the captain wanted to leave the area so fast. He had lost two men KIA and about four wounded. That seemed a reason to stay and kick some VC ass. We wanted to leave also, but we were not leaving without our casualties. Lima Platoon had eight men wounded and two killed.

After we talked to Captain McClure and Colonel Tronsure, we found out they were both ticked off at the captain. Captain McClure said he thought the mechanized unit captain had his orders and that was what he was told to do. I would have changed my orders and adapted to the situation and perhaps saved a few lives. However, in the heat of a battle, no one knows what goes through someone else's mind.

I did know one thing: my guys and I were not going to be charging into any jungle. If Captain McClure had been up there with us, we could have told him the situation, and he would have changed things around and come up with another plan of attack. That's what was so good about our leaders. There was no use in charging into the bamboo; it was a deathtrap.

After the firefight was over, Mike Platoon came out to help us clean up the area. We didn't do too much, as we were worn out from

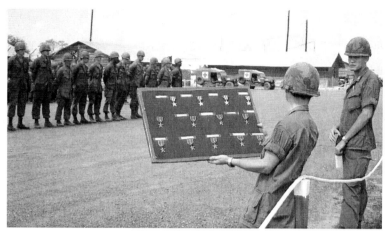

1968. Awards Ceremony for Mike Platoon, C Company.

the fight. Mike Platoon swept through the area looking for any stragglers. I can't remember the exact number of VC that were killed; I believe four or five. We got a few weapons: one RPG launcher, a few AK-47s, a mortar tube and a lot of rounds, and one machine gun. You would always find hand grenades, ammo, paperwork, and VC money from the South and the North. Blood trails were everywhere going into the jungle. We didn't know if they were wounded or dead VC/NVA.

Mech. Patrol with the 11th Cav.

Kenny Gardellis

The 11th Armored Cavs' commander, who is a captain, outranks our lieutenant, so we have to do everything he tells us, which basically amounts to sitting on top of the tanks with machetes in hand to cut any low-hanging branches or vines as we crash through the jungle. The tanks are moving fast, leaving us on top ducking just in time to miss branch after branch. It's almost as though they've forgotten we are up here. We go from jungle to dirt road, back to jungle, and back onto dirt roads. At one point we are on a road and going about 40 mph when the lead APC decides to go into the woods again. Just as it enters the jungle, it is hit by two RPGs that kill the driver and wound another guy. Immediately the other APCs get into position and begin firing into the trees, their 50 caliber guns blazing, cutting through bamboo and tree easily, with wood chips flying. As we hear the distinctive sounds of AK-47s returning fire, Mercer and I take up position behind the tank and hold our fire to conserve ammo. Eventually all the shooting stops, and it is absolutely silent. The air is still, and the cordite smoke is thick

At this point Mercer and I know that it is time to pull back away from the initial contact and call in artillery. However, the mech. captain orders us to get in line shoulder to shoulder and walk into the jungle to get a body count. We try to convince the lieutenant that this is a bad idea, but he is fairly new and has little experience compared to the nine or ten months Mercer and I have been out in the bush. Our lieutenant feels he has to obey orders. The lieutenant marches on into the woods, and Mercer and I hold back to get at the end of the line. I grab a fairly fresh grunt, Strano, and hold him back from his eagerness to follow the lieutenant. I like Strano. He's new but smart and cautious with a lot of courage. He's also from New York, like I am, so I take a liking to him. Later in my tour, while Strano is walking point for me when I'm promoted to sergeant, he takes the daring job of going into a hot enemy campsite after a small firefight and brings out two brand-new AK-47s with factory grease still on them. He really should have received a Bronze Star for his actions and courage.

Back at the site of the APC firefight, Mercer, Strano, and I hook onto the end of the line, barely in sight of the lieutenant, and all hell breaks loose ahead of us. We pulled back out of the bamboo and went back down to where the lieutenant was, and the .50 calibers and AKs go at it again. The VC also had a .30 Caliber gun working for them. When all is said and done and the firefight ceases, we discover that the lieutenant and his radio man have been shot and killed. We load up the bodies onto the APC and head away from the area. We hear the sounds of artillery exploding up in that hot area, but Mercer and I know it is probably too late to get any enemy soldiers. It has taken a half hour from the initial contact for someone to call in artillery. (Interestingly enough, many years later, after successfully surviving his 365 days in Nam, Strano contacted my mother and said to tell me that he made it and thanks for saving his life. Of course, at the time I was just doing what my instinct and many months in the bush told me to do. The fact that he was thankful for me has helped me get through some bad times dealing with PTSD. It made me feel a little better about my Nam experience.)

Attack at the Dong Nai River

Tom Mercer

We left the Song Be area and headed toward Saigon. We were going to guard an engineering company that was building a bridge across the Dong Nai River, not far from Saigon. The bridge was going to be about 700 feet in length.

It was strange the way that Dogface Six had us setting up our NDP. Alpha and Charlie Companies, the Headquarters, and the Recon Platoon were on one side of the river and Bravo Company was on the other side with two companies from the 11th Armored Cav. Bravo Company didn't like being by themselves and away from us across the river.

The reason for the bridge was so the 11th Armored Cavalry could get across the river. The Dong Nai River is one of the largest in South Vietnam. A couple of days later they probably wished they had stayed on the other side.

April 27, 1968. Colonel Tronsrue got a call from another battalion commander saying they had captured some NVA and determined that there were 300-400 NVA and VC heading our way. Thank God not quite that many showed up. That many NVA/VC could have done a lot of damage if they had hit on one side of the NDP. Even though they did pick Charlie Company that night, it wasn't 300 NVA/VC that showed up.

We had plenty of Claymore mines and trip flares that were set up to cover the front of our bunker line. Our ambushes were sent out ¼ mile from the NDP, and our LPs were out, so we were ready for whatever came at us – or so we thought.

What happened next made us worried. The tanks and APCs pulled up beside our bunkers and dropped their guns down horizontal. They stored a lot of "beehive" rounds, and that could only mean that they were expecting something big to happen. When tanks dropped their guns down, like that one did, they were expecting a human wave, when the NVA/VC line up and charge. You must be ready when this happens. I was a little nervous. My time in Vietnam was getting short, and I wanted to leave on my terms, not the enemy's way – in a body bag.

Colonel Tronsrue didn't tell us about all the VC that were headed our way until 42 years later. At the time, I'm really glad he kept that information to himself because that would have been very scary to know. We were expecting a couple hundred VC, and we could handle that. They would have needed more than that in order to whip us. There may have been that many, but they didn't hit on one side at the same time. They were scattered out around the NDP. I think the majority hit on our side, but we did what we had

April 27, 1968. The bridge that was built across the Dong Nai River.

been doing all year – we kicked their asses.

All the officers were running around in a tizzy, making sure everything was set up right and the ammo was distributed out equally, and we had plenty. For some reason, Lieutenant Smart, C Company, told us to have our gas masks ready in case we needed them. I didn't because I had lost mine, so here he comes running in front of some bunkers, which almost got him shot, to give me my gas mask.

Lieutenant Bryon Dennison pulled us squad leaders to the side and briefed us on what was going on and what to expect. When I got back to my squad, I told the men what I had been told. The men were nervous. April had seen a lot of our guys killed or wounded. Everything worried us old-timers, especially with all the new guys we had in Charlie Company. We didn't want them or any of the new guys hurt or killed. But the FNGs had been taught the right way to do things.

Our ambushes were ordered to blow the Claymore mines and head back in. They had been hearing movement all night, as did the

LPs. As they were heading in, the squad leader in charge of the ambush kept saying the "gooks" were right on their heels. The ambush patrol got within 100 yards of the NDP and called in to say the enemy had stopped. They made it back to the NDP along with the four men from the LP. Nothing as bad as having to blow your Claymore mines because the NVA were coming to you. The man who has rear security has the worst job in the world at a time like that.

One LP could not get back in. The NVA/VC came up on them fast, and they did not hear them. They got down as low as they could without moving but were ready to fire, just in case the VC found them. The NVA/VC were looking for them but finally pulled back.

It didn't take long after the ambush and the LP was back in that the mortar rounds started coming in. All you could hear was thump, thump, thump from three or four different areas. They were sending them in as fast as they could drop them down the tubes. It lasted for at least 15 minutes, and then stopped as fast as it started. It's a good thing that the ambush and LPs were back in or they would have been caught in a real bad position. The one LP that was still out finally came back in safely but shaking from their ordeal. Actually, they were better off in front of the NDP since the mortar rounds were hitting inside our perimeter. The NVA and VC were trying to knock out our mortar tubes; they were the main target. Another target was the equipment inside the NDP. There were gasoline tankers, bulldozers, trucks, APCs, tanks and anything that an engineering company would need to build a bridge.

As soon as the mortar rounds stopped coming in, our mortar platoon starting giving it right back to them. The NVA /VC were making their move now. I remember being nervous and all hyped up as the ground attack started. It's a scary situation to be in – you're waiting to start shooting and can't see anything, but you knew they were definitely out there. From the movement and the noise we heard, they were within 30 yards of us. When the NVA /VC got with-

Dong Nai River. April 27, 1968. Destroyed Armored Personnel Carrier.

in 20-25 yards, we could see silhouettes of their bodies running through the tall grass. Once we saw the trip flares going off is when we started giving them hell with small arms and Claymore mines. The Claymore mines were great for something you can't see. The enemy kept coming, and our mortar platoon kept giving them hell. They were right in front of Lima Platoon and were trying to get in badly. Reinforcements came from Mike and November Platoons, and we needed it.

The NVA/VC would hit one side, then move to another side. They kept us real busy for a while. We were definitely killing a lot of them. We were told to always get a good body count. We never knew exactly how many were killed in most of the firefights because they were great at dragging the wounded and dead off into the jungle to hide them.

Michael Shapiro, from Oscar Platoon, was lighting up the sky for us with illumination rounds. We could see real well, but so could they. Between our rifle and mortar fire, all they could do was retreat, but they came right back for more.

I remember a tank and an APC and a fuel tanker. Enemy fire

Swimming in the Dong Nai River.

damaged the tank and blew the APC apart. The top and right side were blown off. Then the gasoline tanker exploded. The heat from the blast melted everything inside the APC. The men didn't have a chance. My bunker was to the left of the APC when it blew up, and so was Kenny Gardellis. It was a terrible sight to witness. I don't understand how our men so close to the explosion survived.

David Ballard and Lon "Smitty" Smith were both hit. Lieutenant Smart was hit in the chest, but all three lived. Three guys in the 11th Armored Cav. were killed in the mortar attack. David Ballard lost the sight in one eye, and Smitty was paralyzed for a while, but partially recovered. We had a lot of guys wounded that night. I can only look back and think about what might have happened if they had attacked in a big human wave like we were expecting.

Captain Phil McClure, our past company commander, was in the rear as S3 for the battalion. He hopped the first chopper out to the field to be with Charlie Company and the men he had once commanded. He was a great company commander, and all the men respected him for being with us during the bad times.

I have to give a lot of credit to Michael Shapiro and the guys in Oscar Platoon. They kept the NVA and VC beaten back from us on the frontline. It seemed like whenever one platoon got in trouble, another platoon always came to the rescue. When the mortars pounded us, our mortar platoon had to be standing up firing our mortars back at the enemy, putting themselves in danger. But they did it every night. Mortar platoon, I'm sure I owe my life to you guys. As always, you did a great job at the river.

After the night was over, we saw the mess that was left – dead VC in front of our bunkers, really close. The way the dead bodies looked made me glad I was on the American's side. They were torn all to pieces. The guns from the APCs and the small arms fire and the M-60 machine guns from Charlie Company tore them to pieces. Body parts were everywhere. The Claymore mines did a great job in helping the body count that the Army wanted so badly to make sure we were winning the war.

Colonel Tronsrue really came through with the right information as he always did. He was and still is a great leader, and I am really honored to have served with him in Vietnam and to call him my friend. His command started around Loc Ninh when he was getting ready to take over the 1st Battalion, 18th Infantry. I was in a lot of battles and firefights under his command, and we always came out looking pretty good. Of course, we lost a few men and that always bothered the colonel, but he made us pick up and continue. He tried not to give us time to think about it too long, knowing how it would affect us. But there is no way we could not think of our fallen comrades. We did think about them and that just gave us the mindset to carry on.

I believe we only stayed there for a few days, and I was definitely ready to leave the area. The only good thing about the river was that we got to go swimming. (Of course, we had guards watching us the time we were in the water.) We were buck naked on our mattresses, but we were swimming. It was like a little R & R, and we had fun.

Firefight at the River

Alan Moniz

I remember the night we were hit at the river while doing securi-ty for the engineers who were building a floating bridge to bring the Armored Cavalry across.

The day had started like any other day, with the usual patrols going out to do recon around our NDP and some of us getting hair-cuts and shaves from one of the local barbers from one of the villages that were allowed inside the perimeter. I recall that the engineers were finishing up on the bridge, and the Cav was getting close to making the crossing. Later in the day word was passed around that that we were going to be hit that night by a large enemy force oper-ating in the area. They wanted us to stop work on the bridge so the Cav could not make the crossing. We cleared our fields of fire and did all the normal things that we did before nightfall in preparation for a possible assault I prepared my position for the night with plen-ty of ammo for my M-60 and everything placed where I could get at it in a hurry. I didn't have ambush patrol, so at least I'd be able to hunker down and be ready.

Time passes very slowly while you're waiting for something to happen and that night was no exception. We thought that we had movement in front of our position a couple of times, but your eyes start playing tricks with you when you are watching so intensely because you are expecting something to happen. So we just kept waiting and watching. I believe it was about midnight when the first trip flare went off, and then a pause before all hell broke loose with green tracers coming in and red and white tracers going out. Even though I was firing my M-60, I recall the sounds of mortar rounds impacting around my position and the deafening sounds of

Dong Nai River. April 28, 1968. Night after mortar attack.

the .50s firing from the armored vehicles.

The Cav arrived around four o'clock in the afternoon and set up a defensive position with APCs and tanks in between our bunkers around the perimeter. I recall seeing an RPG flash by and hitting something off to my left, but I was down in the bunker firing out of the firing port at that time and couldn't see what it hit. But there was a loud explosion, so I knew it wasn't good.

After things quieted down a little, I recall dust-off coming in to take out the wounded even as we were still taking incoming fire and wondering who was hit and how bad. It was a terrifying night that seemed to last forever.

When daybreak finally arrived, we assessed the damages and searched outside of the perimeter for weapons and a body count. Among the dead we found the barber that had given us haircuts and shaves that very morning. We assumed he must have been their scout in preparation for their attack. Fortunately, he was there in the morning before the Cav arrived; so they were caught off guard by the tracks in our perimeter and never had a chance to do the damage that they wanted to. To this day I can still feel the straight razor on my throat that he used to give me a shave and will always remember how from that time forward I never had anyone give me a shave again.

The Dong Nai River

Kenny Gardellis

April 27, 1968. One time Charlie Company is airlifted out of Lai Khe and flown by helicopter to a bridge on a river on which an engineering company with tanks and APCs from the 11th Armored Cav has just done some work. The company is camped out in the open, and we are to dig in and guard the engineers until they are ready to leave the area. They are completing a temporary pontoon bridge to allow other tanks and trucks to cross the river. We wait. Air temperature is above 100, and the river is so close. It looks like it would be easy to get into. Our commanding officers allow us to go to the river as long as we keep an armed guard on alert for safety. So, one platoon at a time, we go down, strip down naked, and jump in for a swim. It is so cool! Actually, it feels more like a cleansing bath than a swim; we are always in desperate need of a bath.

It is so nice and refreshing and relaxing, but my eyes keep looking over to the opposite bank, looking for signs of the enemy in all the thick-green vegetation. I sense we are being watched by VC. Even in our state of relaxation, however, we are always paranoid and vigilant, and we still keep our eyes and ears open. By this point, nine or ten months in, my senses are well honed to the sounds that Vietnam makes during the day and night. I know the sound of ani-

tApril 19, 1968. An M-48 tank from the 11th Armored Cavalry Regiment.

mals as opposed to the footsteps. I can immediately sense an area and size it up. I live every day as a native. I wake up on the jungle floor or rice paddy. I am literally "in" Vietnam.

While guarding the engineers near the river, I meet a nice, talkative guy in the perimeter who is the driver of the fuel truck. We talk into the night and listen to his little portable radio that he puts on his dashboard. Hearing rock 'n' roll in a combat zone was something to remember. We discuss the differences in our jobs over here, and he explains how he sleeps under his truck. Eventually I head to my bunker that has an APC parked next to it, which somehow makes me feel a little safer. We send our LPs out and settle in, hopefully, for a quiet night. It was after midnight when we start getting hit with mortars followed by a ground assault – very intense and deadly. The LP come running in and are looking for a bunker to get into. The guys from the APC, which is now on fire, have jumped into our bunker, so there is no room for the LP. Their APC, which caught fire from a flare, is loaded with ammo, blocks of C-4 explosives, and riot gas canisters. No wonder the Cav. soldiers made a quick exit. The other APCs are fine and are lighting up the place with their awesome firepower. Mike Shapiro and the guys from Oscar Platoon are firing illumination rounds that light up the sky above the battlefield. The APC that is ablaze is only about ten feet from our bunker, and I am exposed as the last man in our bunker. I'm sitting on the steps leading into the hole. My head and back are exposed, and I ask someone in the bunker to pass me a steel pot for my head.

The captain calls in on the radio and tells me to get the men out of that bunker quickly. Every time I tap a guy on the shoulder and tell him to get out and run for his life, there is something exploding in that blazing APC next to us that scares him back down into our bunker. Twenty minutes later the captain informs me that the riot and tear gas canisters are exploding and sending clouds of noxious gas into the perimeter. Many guys have to put their gas masks on. Mine fell off during a river crossing months ago. Luckily, the gas cloud is blowing away, so we don't get any ill effects. However, the captain continues to call and convey the imperativeness of getting my guys out of the bunker before the gas tank explodes on that APC. Shortly thereafter, it happens. As the bunker is filled with the sounds of the guys in there praying, I see the night light up in a searing white-hot explosion. Momentarily I imagine I'm toast as the APC blew up. Huge pieces of the steel came raining down all around us. How was I not hit? They say there are no atheists in a foxhole, and after this long night, I believe that to be correct.

Guys are hit with metal chunks and start running and calling for a medic. It is complete chaos. We still have to get out of our bunker because the bottom half of the tank is still on fire and cooking-off 50 caliber rounds, grenades, and blocks of C4 explosives. One by one I have each guy in the bunker run to the next bunker further away from the burning APC. I'm in constant communication with the captain, who is in the center of our NDP. I've had the radio handset gripped tightly in my hand and held to my ear for a couple of hours communicating with HQ. It is still quite dangerous to make a run for it out of our bunker, but there is no question we absolutely have to get out of there. We wait for lulls in the action then have one man at a time make a mad dash. I am last to go. I take my radio but leave my rifle behind by mistake. I spend the rest of the night in a position, sitting behind some sandbags without a weapon. I keep my eyes open, looking all around, as the flares light up the battlefield.

I listen to the radio constantly and hear medics talking to the captain, informing him just who was hit and who died so far in the nighttime battle. I hear someone call for a medevac. Three good friends of mine are wounded: Smitty, David Ballard, and Tom Christopher. Smitty leaves his M-14 with the captain, and the next morning I go over to my bunker, next to a charred shell that was once a flaming APC. I look around my bunker where everything is burned beyond recognition. I find my M-14 rifle burnt to a crisp, the wooden stock completely gone. All that remains is the steel barrel. Luckily, someone gives me Smitty's M-14. I am so lucky to get that M-14; this one is in better shape than my old one. The M-14 is a bad-ass rifle, and I've learned to depend on it as point man. The next morning Mike Shapiro comes over to my bunker to ask how I am. I simply say, "I'm a believer," referring to the prayers that I believe must have helped us survive the blast. But not everyone is so lucky. During the ground attack that night, our mortar platoon was courageously returning mortar fire against the VC. One incoming mortar or an RPG hit the fuel truck and made a huge explosion. I find out in the morning that my new friend, the truck driver, had been sleeping in his usual spot under his gas truck and died, probably instantly. I only talked to him for a little while. I'll always remember him, our conversation, and the rock 'n' roll music that filled the airwaves earlier that night.

Military Police Duty

Alan Moniz

When I returned to the states after my tour in Vietnam, I was reassigned to Ft. Hamilton, New York to finish my last five months of active duty. Once there I was assigned to the 213th MP Detachment, where I received training as a Military Police Officer. I was assigned to a unit that patrolled and was responsible for the security of Ft. Hamilton in Brooklyn, Ft. Wadsworth across the river on Staten Island, and an airstrip on Long Island.

When I arrived, there were no apartments on or off base supplied by the military that were available to rent, so we had to rent a third-floor attic apartment on Staten Island. The problem with living on Staten Island was that I had to report for duty each day at Ft. Hamilton, which was across the bridge with a toll of $1.00 each way. On my pay, there was no way that I could afford to do that. The platoon sergeant had to improvise by having one of the patrol cars finish up their shift on Staten Island and pick me up there and take me over the bridge to stand formation before my shift would begin. When I finished my shift, I would turn in my weapon and catch a ride with one of the patrols going back over the bridge.

After I got to know some of the guys in the motor pool, they gave me some bridge passes that they acquired somehow so that I'd be able to cross the bridge more often with my own vehicle. Once again I learned to adapt to the situation much the way that I had in Vietnam.

Having purchased a new car while on leave after Vietnam, I was saddled with a car payment along with high rent for the apartment off base, which left me short of cash each month. Needless to say we didn't have a phone. In order to be able to go home to Massachusetts on a two-day pass, I would have to loan my new car out to a friend to use on occasion so that he would leave it full of gas, which would enable me to drive home and back with that tank of gas.

After paying all our bills at the end of the month we would be lucky to have enough left to have a cone of ice cream. So my wife saved every penny she could and near the end of our time in New York she cashed all the change in at the local corner store and gave me ten dollars so that I could go for a pizza and beer with some of the guys at the Enlisted Men's Club on base. She said that she felt bad that I could never go with the guys because we couldn't afford it.

One of our jobs as MPs was to raise and lower the flag at the parade field during the playing of reveille and taps each day. To keep ourselves amused, we would put a beer or soda can in the cannon to see who could get the best distance out of it. This continued until the base commander saw us and put an end to our fun.

Toward the end of my time at Ft. Hamilton, the apartment I was renting ended about two weeks before I was to be discharged. So my wife went home to Massachusetts and I stayed. Now my problem was that I received quarters allowance, so I could not eat in the mess hall or sleep in the barracks. Once again we had to improvise. The guys would take me to the mess hall as a prisoner in handcuffs so that I could eat. One of the guys used to get a kick out of hitting me a little with his nightstick to make it look real. One of the cooks always said you look familiar, but I always said he must be mistaken. As for the barracks, nobody really kept a close watch on that, so I just put my stuff in one of the guy's lockers and slipped in and out pretty much unnoticed.

It was an interesting part of my life that I thought I would share with my Charlie Company family, and I would not change a thing for I believe the experience made me a better person.

The Death of Jack Wayne Oakes

Tom Mercer

June 17, 1968. It was a hot night, quiet and nothing moving around except Lima's Ambush Patrol. It was a weird feeling, but I always felt like that in a free-fire zone. We went out about one mile from the NDP. Lieutenant Dennison had everything logged in with our mortar platoon and with the FO, who called in the big stuff. At least we had a good platoon leader in Lima Platoon at the time. I was in charge of First Squad, which I had been in since arriving in Vietnam in July 1967. I had twelve days left in field and was getting short and nervous about being there.

Tommy Strano was a good man to have in my squad, along with David Gilbert, and Jack Oakes. Oakes had only been in the country for two months and was already carrying the M-60 machine gun. He was learning the "how to's" of Vietnam real fast. Lauren Coleman was the other squad leader on the ambush that night. He had his guys set up on the backside of the ambush. Coleman had been in country since August 1967.

We had 16 men on the ambush. Lima left two squads at the NDP to defend our section. The ambush was to be set up on both sides of a road that was 10-feet wide. The road went through mounds that were 8-feet high and 30-feet long on each side of the road. We divided the men into four four-man positions. We kept one man awake on each position all night. This way we would have four men awake at all times. This was a bad area and the NVA were everywhere.

Our Claymore mines were set up all around our ambush site. We could blast anyone who came in the area, providing we saw him in time. The Claymores by the mounds were set up 45-50 feet away from our positions. At least we had something to keep the back blast from hitting us. The Claymores on the back side of the ambush were out as far as they would go, except two of them. They were behind a stump. This ensured we had protection right by the two positions in the back.

Lieutenant Byron Dennison had been in country for a few months – still an FNG. He was a real good platoon leader and getting a lot of experience under his belt He took over Lima Platoon right after the patrol and ambush when we lost Lieutenant Mellow. He caught on real fast, so that made it easier for all of us grunts to like and get to know him. I've got to say he did a real good job in setting up the ambush that night, of course with the help of us old timers.

We had one M-60 machine gunner on the mound, which was Jack Oakes, and he was a good one. We had plenty of firepower on the backside of the ambush, covering both sides of the road. It was a very good set up and the best way to do it in order to get coverage over the entire area. It was a clear night – at least it wasn't raining – so our visibility was great. So I thought! We set up the ambush thinking the enemy would be heading toward the NDP, but instead they came from the direction of the NDP. It didn't matter; Lieutenant Dennison had the ambush set up great.

Hours had gone by and nothing happened; so we thought it was going to be an uneventful evening. About 4:30 a.m. hell came to visit Lima Platoon. The VC had run into the backside of the ambush and stopped. Why they stopped, who knows? Three of the VC grouped together thinking they were safe from our men back at the NDP, and I know they didn't see us. Lauren Coleman blew his Claymore mine and killed the three where they were. The remaining VC took off running right through our ambush position. As they passed through the mounds, eight of us jumped on the mounds and started firing. We were tearing them up. Once they got 20 yards out, they stopped, got down, and started returning fire because they knew they were about to get their asses kicked. After a while the return firing slowed down. I wasn't sure it was because they had left or they had died. When I looked over the mound, I could see three bodies lying on the road, and they appeared to be dead. We were still firing in that area.

Tommy Strano and Jack Oakes were good friends. Tommy always looked after Oakes, mainly because he was a new man. In Nam you needed to have a good friend that had been there a while to teach you the ropes. Tommy Strano was taught by Kenny Gardellis and me, and Strano turned out to be a good soldier. Another friend of Oakes was Gary Hutto, and they were together a lot. For some reason Gary Hutto was not on the ambush with us. He stayed back at the NDP.

Out of the blue I heard a couple of rounds from an AK-47. That was when Strano started yelling, "Oakes is hit! We need a medic fast!" Tommy Strano jumped right into action and grabbed Oakes and pulled him off the mound. The M-79 thump gunner was now helping Strano with Oakes. It appeared Oakes had been shot in the head, and Tommy was trying to hold everything together. I ran to the mound where Oakes and Strano were when Jack got shot to find out just how bad it was. By this time the medic had arrived and was working on Oakes.

I had to leave, because I had six more men I had to check on to make sure they were okay. I gave Lieutenant Dennison the status of Oakes and that we need a chopper real fast. Lieutenant Dennison called, but they couldn't come out because we had not secured the area yet.

I was starting to lose it. With only 12 days left in the field, I was really nervous. I started yelling at Lieutenant Dennison, and finally he told me to shut up and do my job. I wasn't yelling out of disrespect. I guess I was freaking out. One freak out in a year. I guess I had it coming. There was only one way to secure the area and that

1968. Jack Oakes with machine gun. Tommy Strano with M-16.

was to go out to where the enemy was and get it done. It was a scary situation. We were hoping the VC were dead already.

Lauren Coleman and I and one other guy, whose name I can't remember, headed out of the safe zone. All of us had been in country for a while and that made me feel a lot better. But as we started out to where the enemy was, we started taking fire. The VC were firing from two different positions, but they still didn't see us. This confirmed that they were not dead. Once we pinpointed their location, we did what we had to do and took them out, which took about 10 minutes. We went to all the bodies to make sure they were dead, and we counted a total of five VC. Now the choppers were coming and could land where the VC bodies were lying. It appeared Oakes had died in the arms of Tommy Strano. Oakes was put on the chopper and airlifted out. This was the last time we saw him. Gary Hutto and Short Round (Rick Morrow) were picked to go to Graves Registration to identify the body.

Strano said Oakes was doing what he was supposed to do – looking to see where the VC were and doing his best to kill them. It was a fluke shot that hit him in the head. Strano said Oakes only had his head up just a few inches so he could see what was going on; that's when he got hit. That's the kind of person Jack Oakes was. He wanted to be right in the middle of the battle and wouldn't hide from it. Oakes was a great guy and great soldier to be fighting beside you in a firefight.

Tommy Strano and I have talked about this incident several times in the last year. Strano has gone to counseling for years, and this is the one thing he talks about. It's not the worst ambush we had, but when someone dies in your arms, you take it real hard. It's something that you just can't get out of your mind. Jack Oakes will always be in our memories and with us at our reunions.

Hurry Up and Get Back

Kenny Gardellis

One hot, humid, sweltering day I am walking point with my radioman. I see a suspicious area across the other side of a rice paddy, and so I call my lieutenant and tell him to hold up while I check it out. I have to take it slow because I am using a dike for cover and have mud up to the top of my boots, trying to suck them off my feet. The lieutenant keeps calling me over the radio, trying to get me to speed things up because we're running late for our estimated pickup by the choppers. At one point, in the middle of his persistent badgering over the radio, I half-jokingly tell him, "Sir, if you think you can do a better job at it, then you come up here and walk point yourself." I don't want to walk directly to this suspicious area in an open straight line. I'm trying to sneak up on it. The lieutenant's constant prodding pissess me off, and I start hurrying through the middle of the paddy toward the suspicious

Charlie Company getting ready for a flight out to who knows where.

area. Now the radioman and I are in the middle of the paddy and vulnerable as hell. At this point I'm so mad that I don' really care anymore. I feel as though the VC are in the wood line, and I'm going to be shot. As soon as we get out in the middle of this mucky rice paddy than a light observation helicopter (LOH) hovers right in front of us just 20 feet off the ground. The copilot waves his arms in the air, which means to me: "Go back! Get out of the middle of the paddy!" Obviously they can see the suspicious area, and it looks like trouble to them.

So here I am caught between a rock and a hard place. The lieutenant is yelling at me to go ahead faster, and the LOH guys are telling me to go back. We end up going to the forward toward the wood line where an enemy bunker stands. It is extremely lucky for us that the bunker is empty – except for brass casings from a 30- caliber machine gun that let me know that the VC had been here recently. I'm pissed as hell at the lieutenant for rushing me and nearly walking us to our deaths. Had there been any VC at that bunker, as the LOH and I suspected, we would be dead. I'm angry and exhausted. The radioman and I sit down in the shade of a steam bath-type day and drink some water, wipe the sweat off our faces, and wait for the rest of the company to catch up to us. Eventually the lieutenant comes up and declares that he is, indeed, going to walk point. I am shocked, but say, "Okay," and take up a position with the rest of the company. As the new lieutenant brazenly walks his version of point man. He takes out his map, folds it up, and puts it in his helmet band. "What a dumb move," I think. "The VC will know immediately that he is some sort of high-ranking officer with that map shining in the sun." For 45 minutes or so he moves his way through the open terrain, and all of a sudden I hear, POP. A sniper's bullet rings out loud in the heat of the day. Words gets around fast that the lieutenant has been hit. One shot, in the neck. Not good. They airlift him out, and we never see him again. I'm not sure whether he lived or died. He was so new to our company, I don't remember his name. Given the circumstances, there is only so much you can do. The enemy is unpredictable and hard to find.

Going Home

Tom Mercer

July 5th 1968. Finally my big day was just about here. I was in the final days in Vietnam, and I was ready to leave. I had my uniform dry cleaned, and everything was in order. I had 12 days left in country, and the time was not going by fast enough.

Even though I was excited about going home, I was worried about my friends who would still be in the field in Vietnam. I knew how things could go bad in a heartbeat. We had a lot of new guys, but there were a lot of old timers to help them when times got tough.

Once the day came for me to head to Bien Hoi Airport (July 14), I was nervous again. I kept thinking, "What if the plane got shot down?" After all, we were still in a war zone and in Vietnam airspace. Once the freedom bird took off, it was so quiet you could have heard a pin drop. None of the guys were talking. They just looked out the windows at the jungles below, knowing we had friends down there somewhere. As I looked around, there were a lot of wet eyes. I know mine were from worrying about my friends I left behind.

Finally we heard that word from the pilot, "Guys we are now out of Vietnam airspace – welcome home." It was like an explosion had hit. There was yelling, crying, hugging; we had finally done our time and were out of Vietnam, thank GOD. I said my prayer and went back to celebrating with the others.

We were heading to Ft. Dix, New Jersey. I couldn't wait to get to the USA. Then I would know I was really home. When we landed at Ft. Dix, everybody was excited. As soon as we got off the Freedom Bird, they lined us up like cattle and started doing our paperwork and got us on our way home. Ernest (Smoky) McNeal was the last person that I would see from C Company for another 42 years. I never forgot his name or what he looked like; he was a good friend and a good solider.

July 15, 1968. Once I left New Jersey, I headed to Orlando, Florida. I looked toward the air terminal, and to my surprise there were quite a few people waiting for me. My Aunt Bee came running toward the plane and a guard tried to stop her, but she jumped over a small fence. Next came my mama and dad. The guards just gave up on the idea of stopping everybody. Someone told them that I had just got home from Vietnam, so they eased up just a little.

My mom and dad, my sisters, Sonya and Janice, and of course my little brother, Timmy, were there. He was full of questions about my uniform and wanted to wear my hat. There was nothing but hugging, crying, and a lot of laughter. I think that may have been the happiest day of my life up until my daughter was born.

The ride home was quiet. Well, it didn't last long. Timmy came alive again with questions. It felt so good to be riding in a car and not worrying about getting shot at. We even stopped and got a big hamburger and fries and a large Coke. I felt like I was in heaven.

My dad kept looking at me in the rearview mirror, like he knew something was wrong, but he didn't say anything. I was thinking about my friends that were still back in the jungles of Vietnam. I was hoping they would be okay and stay that way. The bad thing is that you would never know who made it out or who was killed.

I believe my mama cried all the way home; she was so happy. I looked around and saw Sonya, Janice, Timmy, Mom and Dad and knew our family was back together again. That moment brought tears to my eyes.

My 30-day leave didn't last long enough before I had to go to Fort Gordon, Georgia, for the remaining one-and-a-half years. I was assigned to a patrolling, ambush, and the M-16 committee. It was a good assignment. I had been there about five months when I got orders to go back to Vietnam, and that just about killed me. I wasn't ready for that.

The only thing that saved me was the post commander had just pinned my Silver Star on me again, I think to get his name in the paper. He then asked me if I was going to stay in the Army, and I said, "No, sir." He asked why not, so I told him I didn't feel like I was ready to go back to Vietnam just yet. The post commander told me not to worry; I would spend the rest of my time at Fort Gordon, Georgia.

I couldn't believe what I was hearing. I did indeed spend the rest of my time at Fort Gordon. But when my time came to get out, I was gone like a runaway dog. I kind of thought, if I stayed in, I would be going back to Vietnam real fast. So I got out. I have since regretted it for 43 years. But who knows? I could have been killed the first day back in the country.

The next 43 years have gone by fast. Once I got out of the Army,

July 15, 1968. Tom Mercer at airport, home from Vietnam, with his mother and little brother Timmy.

I was heading back home to Apopka, Florida, my hometown. I've had good times and bad times, but thanks to God, Mama, and my wife, Joyce, I'm still alive and doing okay.

Through the first 15 years after Vietnam I've had hair down to my waist, and my beard has been as long as eight inches. I've played in bands and traveled all over. I have written 30 songs and even won some song-writing awards. I have three grandsons, two grand-daughters, and a great grandson and a great granddaughter. I have two daughters and one son. My son-in law (John Walker) is like a son to me. He makes sure I am okay in whatever I do. I know when I get real old, I will be in good hands.

Without Joyce taking care of me the way she does, I would not be here. And finding my friends that I served in Vietnam with topped off everything. So thanks to anybody who had anything to do with me still being alive.

Remembrances of an Operation in the Trapezoid

Samuel K. Downing

October 3-7, 1968. 1st Division G2 had received significant intelligence information to indicate that there was possibly a regiment of NVA operating from the Trapezoid. The Trapezoid was an area characterized by heavy, mostly triple-canopy jungle.

Charlie Company was to be the point/insertion company, and as it turned out, we were on point for the entire five-day operation, except for about half of the fourth day, when the Battalion Recon Platoon under Lieutenant Al Burer was on point. Apparently Charlie Company had a good reputation, and we had, as Division Ready Reaction Force, probably seen more action than many of the other companies. It appeared that the new battalion CO wanted to have as many things working for him as possible going into this potentially treacherous area.

After some delay due to weather and helicopter availability, Charlie Company inserted midday on October 3 as point of a battalion-minus "search-and-destroy" operation (the politically correct terminology was "reconnaissance-in-force") at an LZ about 10 miles west of Lai Khe on the west edge of the Trapezoid. We had a hot LZ with mostly small-arms fire, but artillery and helicopter gunship support pretty well suppressed the bad guys. I don't remember any casualties – at least none that required medevac.

The first night, as we were setting up our initial RON positions, I heard (then saw) all kinds of commotion in the battalion headquarters area. The new battalion CO was apparently concerned about having excellent communications, so his staff was erecting an "antenna farm" of huge antennas sticking up above the jungle canopy and making a lot of noise in the process. About that time the battalion CO called the company commanders in to coordinate night defense – and I requested that Charlie be put on some kind of outpost or ambush duty – I wanted to be as far from those antennas as I could get, because to me they would be nothing but a good aiming point for VC or NVA mortars and any kind of indirect or direct fire – even at night.

The battalion CO didn't buy that, and we had to settle for just setting up a very loose perimeter where we were at least 100 yards from the HQ. It was late by then, and most of the men were able to make only shallow depressions, at best, for protection/firing positions. Sure enough, in the middle of the night we were hit with a mortar attack that seemed to go on for an hour but probably was only about 15 minutes – a lot of rounds lobbed in. And, as one might expect, they were primarily landing in the HQ area in the vicinity of the tall antennas. About eight people in the Headquarters group were hit, but I don't recall any casualties in Charlie Company – maybe just some scratches from shrapnel. Charlie Company had to clear an area for an LZ for medevac. Since it was still probably only 3 or 4 in the morning, and with a pretty tight LZ, I had to "ground guide" the medevac in and all the way down to the LZ using two strobe lights. Then I helped get it back out once the wounded were on board. I felt like I made an easy target for anyone with an interest – kneeling with a bright strobe light in each hand for 3-5 minutes as the chopper came in and then lifted back out. Thankfully, the rest of the night was quiet.

The next morning Charlie Company again moved out as point company, moving generally northeast through the Trapezoid and in the direction of Lai Khe. The day was relatively uneventful, although we ran into some unoccupied but relatively new bunkers and tunnels

– many not even completed. Action was minimal – a few scattered sightings and minimal contact.

On the third day of the operation (October 5), we found intermittent groupings of unoccupied but new bunkers and firing positions, indicating that there was a significant force somewhere in the immediate area. Just after noon, our point men, Mark Krofek and James Hunley, saw a few NVA soldiers moving quickly through the heavy jungle vegetation on the other side of a large bomb crater. Mark Krofek actually almost tripped over an NVA helmet at the same time he spotted the NVA. We initiated contact, which was answered by a very large volume of small arms and RPG fire – we had probably foiled a big ambush. We maneuvered two platoons forward and brought in all kinds of artillery fire, helicopter gunships, and Air Force fighter/bombers for hours.

Even with the excellent fire support, we had trouble moving across the open area to recover several wounded or to maneuver to close with the NVA. Every time we tried to move an element forward, our guys were pinned down by very heavy small arms and RPG fire, and incur additional wounded. During the extended air, gunship, and artillery strikes, there were several secondary explosions in suspected RPG emplacements and a probable mortar location. 1st Lieutenant Robert Schwanber, the artillery FO, did great coordination between our troops identifying targets to the helicopter gunships and through the Forward Air Controller.

About an hour into the fight, the battalion CO ordered Bravo Company and the battalion recon platoon to move forward to reinforce Charlie Company and to assist in moving the action forward. Bravo Company, moving to the south of Charlie Company's position, almost immediately ran into NVA in prepared firing positions and bunkers, and was never able to move forward to Charlie Company's position. 1st Lieutenant Al Burer and elements of the battalion recon platoon managed to move up through moderate fire along the north edge of Charlie Company's position but also was unable to penetrate further into the NVA positions.

Throughout the afternoon we were unable to get a medevac to come in, even to a "relatively safe" small LZ about 150 meters behind the front elements. They refused to fly in with that much firing going on. Finally, late in the afternoon, Colonel Carley, the brigade CO, brought his command and control helicopter in and took out, as I recall, 12 wounded, 1 KIA, and a wounded scout dog. There were probably another 15 guys with minor wounds who didn't want to be medevac'd out. My memory is a very sketchy, and I can only remember a few of the more serious casualties:

KIA: Specialist Smith – rifleman with Mike Platoon

WIA: 1SG Ramiro Ramirez (he got a Distinguished Service Cross for his actions that day and was wounded in the head and chest in several, separate incidents. He was medevac'd to Japan – but was back in VN within 6 months).

SP5 Phillip Nelson – medic – bad gunshot wound in the leg.

1SG Ramirez and SP5 Nelson were both wounded in the first hour of the fight as they moved forward to aid and to evacuate some of the initial wounded. 1SG Ramirez continued to fight for approximately two more hours before he was seriously wounded in the chest.

SSG Billie Smith – bad gut wound, not sure if he made it but was alive and reasonably well when I went to hospital at Long Binh to see some of the guys a few days later.

PFC (or SP4) Feid – stomach and leg wound.

The fight went on from just after noon until after dark. At about 8 p.m. the battalion CO had us back up, away from the area of contact, and help form a battalion defensive perimeter. This consolidation went off without incident, and we had no enemy contact during the night.

Based upon specific recollections of firing positions, RPG positions, and spotters/snipers in the trees that were taken out, we killed about 15-20 NVA, with a real likelihood of several more. We were unable to get a sweep forward to check the area until the next morning.

Medic Brian Montgomery, Charlie Company's lead medic, had been sick and at brigade base camp, but when he heard about the fight and Medic Phillip Nelson being wounded, he caught a ride on a helicopter bringing in ammo and water, and stayed with us for the remainder of the operation.

The next morning, October 6, we swept through the area of the previous day's battle. We found some damaged NVA equipment, a lot of destroyed bunkers and firing positions, many drag marks, and other signs that there had been many casualties among the NVA troops.

The recon platoon started out that day on point and continued until early afternoon – but they kept running into bunkers, tunnels, and sniper fire, so Charlie Company was continually moving forward to help them clear out the resistance. Midday the battalion CO told us to take point again and informed me that we needed to be at a designated resupply point by approximately 1800 hours for a major resupply of food, ammo, and water. He told me he wanted to ensure that we were there on time. As we moved forward, we continued to have sporadic, light contact – but there was no question that there was still a significant NVA force still in the area.

After about an hour of relatively slow forward progress, the battalion CO instructed me to move Charlie Company (leading the battalion) down a pretty large jungle trail. From my military education about jungle warfare, rule # 1 of jungle operations is STAY OFF THE TRAILS. In an attempt to avoid moving my company down the trail, and, in my opinion, into almost certain and unnecessary danger, I had an ongoing discussion over the radio with the battalion CO for over an hour while we kept moving through the bush. I tried every trick I knew, including pretending not to hear him because of radio problems. Finally, he gave me a direct order to move the company out on the trail and get moving toward the resupply point. I responded that if Charlie Company was to go out on that trail, it wouldn't be under my command. Not good form, but right. Anyhow, thank God for good communication, and a lot of luck, because Colonel Carley (brigade CO) happened to be in his helicopter nearby, and he overheard the discussion. He broke in and told the battalion CO to me proceed as I was doing.

Colonel Carley's intervention was a good thing because within a few hundred yards, we ran into the flank of an ambush that had been set up along the trail – no telling how many guys we would have lost in that one. As it was, Charlie soldiers only received a few minor injuries. Most of the NVA – surprised at being hit on the flank – cut and ran. We cleared through the ambush area within about an hour, finding positions for at least a reinforced platoon that had been waiting in the ambush position.

We ultimately arrived at our resupply point, about an hour late but in adequate time to receive the resupply. As the resupply was being completed, Colonel Carley set his helicopter down and had the battalion CO come out and talk to him. Within a few minutes, the battalion CO got in the helicopter with Colonel Carley. Colonel Carley had relieved him of his command on the spot.

The next day, October 7, the last day of the operation, consisted of company-size patrols in a relatively limited area and was mostly uneventful. We were lifted out by late afternoon. I think Charlie Company's total losses for the operation were 1 KIA, probably 30-35 WIA, but only about 12 medevac'd. By comparison, a good friend of mine was a company commander in 1/26th Infantry, 1st Division, and they went into the exact same area several weeks later. They had over 30 KIA and 30-40 seriously WIA and medevac'd. We were either lucky, or good, or both. I personally think I was blessed by having some very good officers and NCOs. In addition, the men in the company were pretty well seasoned and worked well together.

Hutto Trapezoid (Diary Entries)

Gary Hutto

October 1, 1968. We were getting ready to move out on a 4-day operation. We were out on the trail and a small helicopter about 800 meters spotted VC from where we landed on our LZ. Now we have to move into the area where the chopper spotted them. The point element of our patrol was checking out a freshly used bunker complex about 200 meters.

October 2, 1968. We were late setting up our ambush site. We were late getting there, and everything was disorganized. The VC mortared our ambush site at 4:30 a.m., injuring 15 and killing one of our men.

October 3, 1968. This morning they found a VC in a bunker. They tried to get him out, but the VC threw a grenade, and two of our men were hit by shrapnel in the arms. Medevac choppers dusted off these men.

We then set up near another VC base camp. We found rice, clothing, and all kinds of documents, which were captured. The artillery pounded the area to keep the VC away; we also had a helicopter gunship above us just circling and waiting.

October 4, 1968. About 10 minutes ago the gunship opened up on some movement. They were firing rockets and mini-guns at the movement.

I'm back and it's about 2:00 p.m. and another man has been killed. He was in the point element with a recon group. He was shot up pretty bad. They are taking him back for the dust-off now.

It is about 4:00 p.m. now and Bravo 1/18 is blowing up the base camp now. Word just came down that we are moving out.

October 5, 1968. Well here it is 8:00 a.m. in the morning. Last night was pretty quiet, only once we had to get in our foxholes when we heard the dreaded sound of VC mortars. The foxholes took about three hours to dig, but I'm glad we had them. We set up in the ambush site. We cleared a spot big enough to set down a helicopter to resupply us with rations and water. We will move out in one hour or so I think.

This place where we are set up was bombed out about a week ago with 500 lbs. bombs full of napalm, and they left a crater about 30 ft. deep and 40 ft. wide at the top. It blew down trees the size of telephone poles and cleared brush like a big bulldozer. The trees left standing have big holes and big cuts from shrapnel from the big bombs.

We got a lot of action yesterday about 1:30 p.m. We shot at what we thought were three VC, but there was a hell of lot more than three. They pinned us down for about five hours. They hit us with everything they had, including RPGs and automatic weapons.

Shrapnel hit me. We had 25 wounded by RPG shrapnel; five people hit by rifle fire. The guys hit by rifle fire are at the hospital in Long Binh. The first one got eight men, including myself. Everybody ran out of ammunition. They tried to bring in more supplies, but the VC shot at the choppers with intense rifle fire.

I was never so scared in my life. My squad leader took my machine gun and left his jammed M-16 for me (Herman [Tex] Vest) was my squad leader.

When shrapnel hit us, my assistant gunner (Wayne Thacker) got hit by a lot more shrapnel than me. He was from Kentucky. Wayne got blood all over my machine gun. The shrapnel I got went almost completely through my arm, but it hit the bone. My arm is stiff because of the shrapnel.

I did some praying while this firefight played out. I think it helped, because I'm here to write this story as seen by an infantry soldier.

Patrol in October

Mark Krofek

It was another day in paradise and, as usual, I was on point. So far it was a quiet day, no snipers as of yet. The past few days we had been getting lots of sniper fire and some mortar rounds being fired at our NDP. As we left the NDP, I was feeling a little nervous. Why I don't know, but I would find out later that day.

The patrol was going to be Bravo and Charlie Company. This was a big patrol, so somebody knew something. But we grunts didn't know anything about it at all. As we moved along, I was still on point, which I had been doing a lot of lately. I could see signs of a lot of NVA movement, and this started to get me a little worried. Still, nobody told us anything except to keep our eyes open and be careful. Those few words can get anybody worried.

We were working the areas six miles west of Lai Khe on search-and-destroy missions, which sometimes got real bad. We had a lot of backup with the firepower of the 1st Battalion, 7th Artillery and the Air Force tactical strikes if needed.

This area had been hit hard with artillery and air support, which meant one thing: lots of NVA. Captain Sam Downing had everything under control. He was the type of CO who cared for his men and not getting anybody hurt or killed was his main goal – besides taking care of the NVA.

I moved along at a slow pace when I came up to a big bomb crater. It would make a good foxhole, I would find out later. I was looking around to see if anything was happening when all at once three NVA took off running to the right of a worn trail I spotted in front of our position. I opened up with my M-14 and killed them, which would be the first shots to be fired in a five-hour firefight. I had two good friends in the foxhole, and they were both wounded. One of my friends in the foxhole was killed three weeks later in a firefight.

The main part of the firefight was going on around the bomb crater. The VC were firing RPGs and lots of AK-47s in our direction. Captain Sam Downing was calling in fire support and a lot of it. At one time he was yelling, "Any closer, you'll be hitting us." But they had to be close in order to do the damage and keep the VC off of us. As the fighting got worse, the battalion commander called up Bravo Company to help us out.

As soon as B Company started toward us, they rain into their own problems. They came upon a large bunker complex full of VC gear and food. One VC got up and was brushing his teeth and got the hell out of there. There was also a block of ice sitting on a bunker. It must have been the same VC we were fighting that were in those bunkers.

I fired so many rounds that I had to get ammo from the machine gunner to use in my M-14, but he had plenty. Five hours is a long time to be in a firefight, but as I look back, it only seems like a few minutes. We were resupplied a couple of times. Finally, it was over. I never thought it would end. We had a few men killed and a lot of men wounded. A few other men and I were decorated for the action that day.

Not too long after the bad ambush, we were back at Lai Khe – or the Holiday Inn, as we called it. We were still in the month of November. The NVA were really mortaring us bad day and night. Then the NVA sent some sappers into the NDP to try and get through the wire we had strung out. They would set up booby-traps all around the area, and those would do the damage if set off.

The night they tried to get in, I was on bunker duty, and it was about four a.m. When the mortars started coming in, I was yelling, "Incoming! Incoming!" as loud as I could. Everybody was hitting the holes and anywhere else they could find cover. Well, my luck ran out on this night. I finally got hit and real bad. Once they got me back and fixed me up, they sent me back to the free world.

It was an honor to have served with the Big Red One, the great 1/18 Charlie Company, Mike Platoon. These men were like brothers to me, and now they are back in my life as my brothers once again. I'm sure in some way they helped me make it back home alive. God bless the men of Charlie Company, Mike Platoon.

My Memories

Robert T. Mueller

I was drafted in the Army in 1968 and somewhat concerned as I was married and a war was going on in Vietnam. I also was established with a good financial analyst job and older than most draftees at age 25.

I spent one solid year in the U.S. taking all the training I could to avoid going overseas. Paris Peace talks were taking place and I had hoped for settlement to avoid going overseas. Bottom line, one year of training, then orders to Vietnam.

After a short leave and emotional goodbyes with wife and parents, I was on a plane to Washington state then Alaska and ultimately Cam Ranh Bay. From there I was assigned to "C" – 1st of the 18th. With all the training I had in the States, I was an E-5 sergeant and wound up a platoon sergeant operating out of a firebase called "Venable Heights."

Several events stand out to me while with C Company. We went on many search-and-destroy missions and many night ambushes. I remember one search mission we went in a village, and I came upon a house with a thatched-roof porch. As I came close, the lady of the house came out to offer me food. I am 6' 3" tall and the antenna on my helmet radio got caught on her thatched porch and collapsed the corner of her porch. Talk about an angry Vietnamese lady going after me with her little, short broom to chase me away with my M-16, Claymores, and grenades. Obviously I fixed her porch before we continued on.

Another memory was we would always on night ambushes try and set up in cemeteries as they provide two advantages: (1) short walls that provided protection; and (2), during heavy rain the mounds where bodies were buried provided a spot you could lay down and stay reasonably dry when not on guard. I found sleeping in a cemetery a little unusual with buried Vietnamese helping keep us safe and dry.

In July 1969 I met and talked with President Nixon prior to going on a night ambush. Approximately one week prior to this visit, our unit was brought into Di An from the field. It was very nice as we had a clean change of clothes, some rest, and hot food. Just prior to going on ambush, our unit was told we would meet the president. All our grenades were locked up and shortly afterward he mingled with the troops. After he left, choppers arrived, and that night it was back to normal on ambush in the jungle.

After approximately six months in the field, I had an opportunity to get a job in Di An as an instructor at Danger U. I went to Di An and had one night to study a radio class and do a presentation to an officer the next day. I got the job and taught classes for the rest of my tour: radio, mine and booby-traps, gas chamber, etc. At this time I was attached to headquarters. At the end of my tour the 1st was to be deployed out of Vietnam, and I was assigned to help a captain to catalog and ultimately escort captured weapons back to the First Division Museum in Wheaton, Illinois. The museum in Di An was a one-room area that new recruits were taken through and shown to familiarize them of enemy weapons, tactics, clothes, etc. I don't remember how many conexes we loaded, but I think it was three.

Something I will never forget is the flight back to the U.S. It was in an Air Force transport, and I sat next to the conexes in a jumpsuit. Also being shipped were several draped caskets of U.S. soldiers being sent home to their families. My emotional range was very high about going home, but low when thinking about the sacrifice these men made and how their families would grieve. Needless to say, it was a long flight and a lot of time for me to reflect on my situation versus theirs.

Our flight went to Alaska and then Texas. The conexes were then shipped by truck where I turned them over to the museum in Wheaton.

Lauren Coleman

Tom Mercer

When I first met Coleman, I could immediately tell he was a quiet, shy type of guy. He didn't say too much and when he did talk, he was really laidback. He did not get in any hurry unless someone was shooting at him. I did not see Coleman as someone who should be in the field, but he surprised me. I also remembered that he was drafted, so he had no choice.

Coleman ended up being one of my best friends. I could always count on him watching my back, as I did his. We were in quite a few firefights together like the Battle of Loc Ninh, the Tet Offensive, the attack at the River, the Power Plant in Saigon, Rocket Day, and the Mechanized Unit Patrol. I could go on and on.

I remember Coleman running from cover to cover following the platoon leader or squad leader. He would hit the ground and all you could hear was a big "Ugh." That radio was a big load, and he had to be firing along with the rest of us. That made it really hard for him, especially in the jungle, but he did it.

When we went to Quan Loi, we would always go to where they had hot showers. We would get our clean fatigues and leave our old ones, which were full of dirt and mud. Once we headed back to the bunkers, within ten minutes we would be dirty and dusty from the Quan Loi clay, which didn't come off real easy, and Coleman always had to be clean.

When I starting getting our first reunion together, Coleman was the second person I found after 42 years. It felt great to hear his voice after all these years and relieved he made it back home safe.

Jimmy Dossett

Lynn (Dossett) Maly

I was a senior in high school, a month shy of my 18th birthday, and had a steady boyfriend. My best friend, Peggy, was engaged to George Dossett, who was stationed in Germany at the time. George told Peggy that his older brother, James, was home on leave, having just left Germany and headed for Vietnam. Jim had taken the leave so he could escort his fiancée to the Senior Prom, but when he got home (around the second week of May) he found out that she already had a date, which she refused to break. George told Peggy that Jim was really bummed out about this. Peggy had never met James, so on a lark, she asked me if it was all right to call and invite him over to my house so they could meet and play pool on my dad's table. Sounded okay to me!

She called and talked to him for a few minutes, then handed the phone to me and asked me to invite him since it was "my" house. James had the sweetest southern drawl I'd ever heard. I made the invitation, and he accepted. About a half hour later he parked his car at the curb and started walking up the driveway, and I think I fell head over heels in love with him the very second I laid eyes on him.

He knew I had a boyfriend. I knew he had a fiancée. None of it mattered. We started seeing each other and were talking marriage after our second meeting. He asked me to marry him, and I accept-

1967. Jim Dossett with wife, Lynn (Dossett) Maly, before Vietnam.

ed. But first we had to get our personal lives in order. He broke it off with his fiancée, and I got my class ring back when I returned my boyfriend's ring. Then he started pushing to get married before he left for Nam. Biggest reason: so he could start sending an allotment to me to put into savings, so we would have a down payment for a house as soon as he got back. My mother was having a hissy fit. She sat up one night into the wee hours of the morning talking until she was blue in the face about waiting until he got done with his year overseas. We could get to know each other so much better in that year, and if we still felt the same way after that, then we could have a proper wedding ceremony. The next day we went to the courthouse and got our marriage license. Jimmy's mom had to sign for him because Illinois law at the time was that a man had to be 21 before he could get married and a woman 18 (which I had just turned). Jimmy's mom got her church minister to perform the nuptials at his home. We were married June 20th and Jimmy left for Fort Ord on June 23rd.

We had plans and dreams. We both wanted five kids. He was going to go back to his old job at the filling station where he had worked before enlisting and eventually planned on buying the place from the owner. We were going to live in Batavia, Illinois, near his folks. He thought he was going to get a motorcycle when he got home. This could have been the cause of our first real fight, but I decided to fight that battle when I got to it.

Back in 1967 I did not even remotely known about what was happening in Vietnam. There were whispers of war, but being 18 and not really aware of international conflicts, I didn't think anything more of Jimmy going to Vietnam then if he had told me he was going to serve another year in Germany. I really didn't know there was any danger. He promised to come back, take care of me, and love me for the rest of our lives. He promised never to volunteer for anything riskier then KP duty. The day before he left, we had a serious talk. If anything should happen to him, he wanted to be buried under a tree. He would rather die then come back missing an arm or a leg. That's when he started to scare me. He was actually going into a combat area.

When he shipped out, he wrote to me every day. I don't think he missed a single one. His letters were sketchy. Once he told me that the South Vietnamese he fought alongside during the day became part of the guerrilla force that turned against him in the dark. He complained about guys sleeping on guard duty. That was about all of the war he was willing to share with me. He lost a stripe (for some reason that I have never known) and was desperately trying to earn it back before he had to tell me. That may explain why he volunteered to walk point on that fateful day. He started asking us to send him cigarettes in our care packages. I couldn't understand why because he had been buying them himself up to this point. Had I known he'd been busted, it would have made a lot more sense. So when the Casualty Officer walked into my place of employment, had my boss call me into a back room, and read to me from a piece of paper in his trembling hands that Private James E. Dossett had been killed, I kept insisting that they had the wrong person – my husband was a Spec. 4.

According to our local newspaper, Jim was the first war casualty from our area. It took a week to get the body back to Illinois, and it was badly degraded by then. The family had a private viewing. Afterwards, we closed the coffin, draped it with the American flag, and put Jim's service picture on top. Jimmy was laid to rest under a beautiful, mature oak tree at Mt. Olivet Cemetery in Aurora, Illinois on Friday, the 13th of October, 1967. After I remarried, James's mother had him moved to River Hills Memorial Park in Batavia, Illinois, where he is at rest today.

My current husband (a Vietnam vet with 27 months in country) and I have been to Washington, D.C. to see the Vietnam Memorial Wall. I have a rubbing of Jim's name – Panel 27E Row 52. We have seen the Traveling Wall on several occasions as well. I also frequent the "Virtual Wall" at the Vietnam Veterans Memorial Fund site on the internet and invite you to go there and type in Jim's name and leave a remembrance.

A fellow vet, Steve Shewalter, became interested in Jim's heroic actions several years ago and set out trying to get him awarded the Medal of Honor posthumously. Unfortunately, Steve has since experienced serious medical problems and is no longer able to pursue this endeavor.

Jim's parents passed away in the late 1990s, as did his brother, George. His brother, Jerry, still survives, and we remain in touch.

Kim, My Big Brother

Daleen (Deeter) Keithley

My brother, Kim, was not quite four years older than me. He was my big brother, and I looked up to him, but I knew he was not a saint. I always thought he felt I was his snot-nosed, pain-in-the-butt little sister, but we were close. Kim and I grew up on the south side of Grand Lake – St. Marys in Mercer County, Ohio, at the Maples Landing and had a great life fishing, hunting, trapping, swimming, boating, ice skating, snowball fighting and enjoying life as kids do.

I remember the day he left for the service. Dad, Mom, and I took him to the courthouse over in Celina and stayed with him until the Greyhound bus took him away. While he was in basic training and advanced training, we lived for his letters, phone calls, and his three-day passes when he would come home to visit us, his girl-friend, and his friends who were at home at that time.

The day we received word he was going to Vietnam was not a very good one for us, but we pulled up our bootstraps and accepted it. He came home on leave before he went, and he, Dad, and Mom went duck hunting together every day and talked about many things. He tried many times to tell them he didn't think he would make it back, but Mom would not let him talk about it. Finally, Dad told Mom to let the boy talk, and he told them about the $10,000.00 life insurance policy he had taken out on himself and what he wanted them to do with the money if he didn't make it back.

The night before we were to take him to the airport, he, his girl-friend, Jane, and his friends who were around at the time went out. The next day Dad, Mom, Jane, and I took him to the Dayton Airport. Dad and Mom didn't talk much, and Kim and Jane were kissing and

Kim Deeter and his mom, dad, and sister.

hugging all the way there. I kept looking out the window because I was so embarrassed having to sit beside Kim and Jane while they were making out with Dad and Mom right there in the front seat. Anyway, we made it to the Dayton Airport and went to his departure terminal. He said goodbye to my mom first. They hugged. She kissed him goodbye, crying the whole time, then went to a window continuing to cry. Next was Jane. He hugged and kissed her goodbye, and she too was crying. Jane then went over to my mom. They were hugging and crying together. Dad shook his hand and told him to take care of himself, to come home alive to us, and gave him the knife he carried all the way through WWII, telling Kim it helped him through a few tough times. Then my time came. He came over, hugged me, and I started to cry. He looked down and said, "Not you too," kissed my cheek, turned, and walked straight out to the airplane and never looked back. Dad and I walked to another window, watched him get on the plane, saw him wave to us from a window, and the plane left. That was the last time I saw my brother alive; he was 20 and I was 16.

Over the months that passed, we again lived for his letters,

1968. Kim Deeter on patrol.

tapes, and a few phone calls. We wrote back, made tapes for him, and sent a lot of care packages. I took all his letters and tapes to my World History class at Celina High School and read his letters to my classmates, plus we listened to his tapes.

I remember getting off the school bus on April 23, 1968 and saw my dad walking around outside. I yelled to him and asked if he was playing hooky from work. The closer I got, I saw that he was crying. I asked if something had happened to our dog, Heidi. He told me no, that Kim had been killed in Vietnam. At first I did not believe him, but I had never before seen my dad crying; so I ran into the house. Several of our neighbors were sitting with my mom, hugging her, and she was crying like I had never seen her cry before. I ran to my bedroom thinking that this can't be true. Mom came into my bedroom and sat on the side of my bed and told me it was true, that a Lieutenant Colonel Rupport had come to the house today and told them Kim was killed on April the 19th. I cried for about an hour, got out of bed, and walking around the landing for a while still thinking that this can't be true and waited for all the neighbors to leave. I then went back into the house, and Dad and Mom told me everything Lieutenant Colonel Rupport had told them.

Dad also told me that he and Mom argued with Lieutenant Colonel Rupport for about an hour that Kim couldn't have been killed. Dad looked at Mom and said I know how he was wounded. After he thought about it, the dates matched up. The day Kim was killed, which would have been in the middle of the night for us, Dad said he woke up and saw Kim standing at the end of his and Mom's bed. Kim looked at Dad and said, "Oh my God, Dad, I am going to die." Dad said he broke out in a cold sweat, blinked his eye, and Kim was gone. Dad was shocked at first and thought he must have been dreaming, settled himself down, and he went back to sleep. When Kim's body came home, his wounds were just as Dad had seen them.

I went to school the next day because I could not stand to see my parents crying and walking in a daze. Going to school wasn't much better. Everyone was coming up and hugging me and saying they were sorry about Kim's death, which would make me start crying again.

We asked if a good friend of Kim's (Dean Armour) could be the body escort since he was still in the service. Lieutenant Colonel Rupport made it happen. After Kim's body came to the funeral home in Celina, Dean came out to the house and told us that it was him – our hopes that it wasn't him were destroyed. We viewed his body in Celina, Ohio, on April 29th; again in Dayton, Ohio, on April 30st; and buried him on May 1st in Centerville, Ohio.

November of 1968 was a tough month for us. Kim would have been coming home, – his year in Vietnam would have been up. Dad, Mom, and I did the best we could to get through Thanksgiving, Christmas, and New Years at all the family parties. It wasn't quite the same as we knew that Kim would never be with us again. But time does ease the pain.

For many years my family hoped and prayed for a phone call from a GI who was with Kim while he was in Vietnam. In 1994 the first phone call came. Rick Marrow from Pennsylvania (now living in Texas) called and talked to Mom. Then about a week later a letter from Bob Quimby from California came stating that he had some pictures of Kim from Vietnam and would we like to have copies. We now have two new friends because of Kim. We were able to reacquaint Rick and Bob, and they have visited Mom and me several times. Mom, my family, and I enjoyed listening to them talk about what it was like while they were in Vietnam, seeing their pictures, and hearing their memories of Kim. About two years later a letter from Joe Boland from Illinois came with a picture of Kim. I called him – now another new friend – and we were able to reacquaint Joe with Rick and Bob. When the Traveling Vietnam Wall came in September 2000 to nearby Coldwater, Ohio, the three of them came for a visit, met some of Kim's hometown friends, and shared more memories. Still more people became friends because of Kim. I stay in touch with Rick, Bob, and Joe by email, and after last year's Charlie Company reunion, I now have two more email friends because of Kim: Kenny Gardellis and David Simpson Jr.

Kim's death had a major impact on my family, but we never got angry because Dad believed and raised us that it was an honor to serve your country. Sure, Kim really did not want to go into the service or to Vietnam, but he had a job to do. He did it, and we were all so very proud of him. Kim's death did not make us angry at God; it made us closer to God and better Christians. I look at Kim's death as only a temporary separation from him. Because I am a Christian, I know I will spend all eternity catching up on old times and make new ones with him.

The Death of Fritz

Rick Morrow

The day we lost Fritz the scout dog was a bad day for all of us. Fritz had been on a lot of patrols with C Company. He was part of the family and a great asset. We were at an NDP outside of Lai Khe. We were on patrol doing a lot of search and destroy, and we were finding a lot of bunker systems. We found a few VC base camps where nobody had been for a few days.

As we set up for the night getting ready for what the VC may give us, we started getting some AK-47 fire into our position, then a few RPG rounds. One exploded close to where Fritz was, and he was hit with a big piece of shrapnel that killed him instantly. We were all in shock, and no words could describe how we felt at that moment.

This is the only causality from a RPG that I can remember. It was like losing a good buddy in the field. After Fritz was killed, nothing else hap-

January 1968. Rick Morrow.

pened that night, but that was enough. All the guys were really torn up about it. Fritz had been with us for a while, way before Rocket Day, April 10, 1968, when he was working with Lima Platoon, 1st Squad. He saved a lot of lives that day. But the war went on, and as

233

usual, we had to get it back together and fight on. Fritz, you will always be in our memories and hearts. You were a true comrade in the war and saved a lot of our guys' lives just by doing your job – sniffing – smelling – stopping – not moving – growling – and, at times, getting mad.

A Story about my 1/18th Charlie Company Soldier, Tom Mercer

Joyce Mercer

I'll begin this story with a little history of Tom and myself. We met over 23 years ago and have been married for 22 years. Tom is a very caring and wonderful man, always putting everyone before himself. He's almost too nice and polite to a fault I know he wouldn't think twice if he had to risk his life for me or any member of our family or friend, for that matter. I consider myself very lucky to have found a wonderful man to spend the rest of my life with.

Prior to the first Charlie Company reunion, Tom never spoke about his experiences in Vietnam. If I asked a question, he would respond with a short answer, and it didn't take me long to figure out not to bring up the subject. During the course of our marriage there are little things Tom does that I couldn't quite understand. I have since found out most of it is due to Vietnam. For instance, I noticed Tom doesn't like going to bed until he is absolutely exhausted and usually he falls asleep in his chair. He has a terrible time sleeping and has to take sleeping medication most of the time. If it were up to him, he would stay up all night and sleep during the day. Tom is also very fidgety while driving or sitting. You can never ease up on him and scare him because you're likely to

Group photo from the first Charlie Company Reunion in Gatlinburg, Tennessee in 2009.

get your head knocked off. Tom has a lot of nightmares and sometimes will talk in his sleep and call me the "general!" I don't know if he's dreaming about Vietnam or dreaming about the present. There are times he'll be very quiet, like he's in real deep thought about something, and I know he's thinking about Vietnam. I could go on and on. After talking to some of the other veterans' wives, I have come to find out that many of these men have the same Vietnam symptoms. It's so sad.

Unfortunately, I have to admit, I really didn't know what Vietnam was all about. At the time I was 18 years old and a working single mother taking care of my baby daughter. I didn't have time to stop and think about world affairs because it was all I could do to be a single parent. I didn't realize how bad and senseless this war was until many years of being married to Tom. The realization for me actually began by visiting the Traveling Wall in September 2008.

For many years, the thought of Tom finding the men he served with in Vietnam was a farfetched idea to him. Little did he know, a visit to the Traveling Wall would be the beginning of the first reunion for the 1/18th Charlie Company.

This all began with an article I had read in our local newspaper about the Wall coming to our town in late September 2008. I mentioned it to Tom because I knew he always wanted to visit the real one in Washington, D.C. Tom was so excited and called our best friends, Jim and Carol Gavagan, to join us. I distinctly remember it was a warm evening and about 40 people were walking around. Tom immediately went to ask an attendant how to locate some of his friends who had died. Tom walked to the Wall with all of us by his side. Within a few minutes, he discovered his best friend's name, Jerry Tucker, and immediately broke down crying. We all felt his emotion and followed with our tears. After walking for a while and looking at names, it was time for the Memorial Service to begin for the fallen soldiers. Neither of us had attended one of these services and didn't know what to expect. While we were sitting, waiting for the service to start, there was a cute little boy with his mother, handing out key chains made with beads consisting of the South Vietnamese colors. He handed them out to all the men waiting for the service. Later we learned that this little boy had lost his father in the Iraq War several months earlier.

That evening, on the way home, Tom mentioned he would like to find some of the men he served with. Little did I know this was going to be his mission for the rest of his life!

As soon as we arrived home, Tom turned on the computer and started. Luckily, we have a program for our business that can locate people, provided we have good information. The first person he looked for was Kenny Gardellis, a young man who became his best friend in Vietnam in 1967-1968. Tom never knew whether Kenny made it home alive. When Tom left Vietnam for home, Kenny was on R & R. He never knew whether any of the young men he served with made it home alive. I know this weighed on him all these years, but he never talked about it.

As luck would have it, Tom found an address and phone number for Kenny. He was so excited and almost speechless. Ironically, Kenny lives in a little town outside Asheville, NC, and only 1 ½ hours away from us. Tom immediately picked up the telephone and dialed his number. I had to go to another room so Tom couldn't see my tears of joy. Kenny's girlfriend, Traci, answered the phone, and I heard him ask to speak to Kenny. He explained to Traci that he served in Vietnam with him. Kenny got on the phone, and Tom was so excited and couldn't believe he was actually talking to him after all these years! I remained in the other room until I composed myself. I could hear plans being made for all of us to meet at a restaurant in two days, which couldn't come fast enough for Tom.

The initial meeting with Kenny and Traci (who have since married) was like something out of the movies. I can remember Tom was so nervous and couldn't wait to see him. Once we arrived at the restaurant, he immediately recognized Kenny, and they couldn't stop hugging each other. Although over 40 years had passed, it wasn't hard for Tom to know it was his best friend from Vietnam. We all tried to hold back the tears. That was the beginning of a 1/18th Charlie Company reunion in the making. Tom told Kenny of his idea of trying to find the other guys, and Kenny, with no hesitation, offered his help. When we left the restaurant that evening, Tom couldn't stop talking about how happy he was to finally find his best friend from Vietnam. I was so happy for him! Once home, Tom turned on the computer and spent endless hours trying to find the men he served with. There were also many hours on the telephone. Every time Tom found someone, he called that person and immediately called Kenny. After several months the reunion date and location were set.

Another man Tom served with, Doug Goddard, lives about 45 minutes from us. Doug and his wife, Caroline, also offered their help with the first reunion. The date was to be April 30-May 2, 2009 in Gatlinburg, Tennessee. Everyone put in a lot of time and effort to make this first reunion a huge success. I believe there were 38 Charlie Company members who attended; many brought their wives or girl-

1967. Chair covers from Charlie Company mess hall.

friends. As the reunion approached, Tom was extremely nervous. Although he had spoken to all of these men by telephone, he would soon be seeing them for the first time in over 40 years. We could hardly wait until we arrived at the hotel. Doug and Caroline did a great job preparing the name badges, which helped tremendously. Doug actually had a picture of each soldier when they were in Vietnam along with his name beside the photo. The reunion was a huge success and a lot of tears were shed. I realized all the hard work and effort was worth every minute. I also realized how much people thought of my husband and what a good, genuine person he is.

After the reunion, Tom told me he would like to write a book on his experiences in Vietnam. My immediate thought was "Oh no, here we go again." Once we arrived home, Tom was back on the computer typing away. Over several months, until wee hours of the night, he had the idea not only to write a book, but to include his Army buddies as well. Tom contacted the other men, and most of them seemed interested in writing of their experiences. It took well over a year for Tom to compile all the stories, form this book, and mail it to the First Division Museum in Wheaton, Illinois, for editing and publishing. I have to admit, after the first reunion, I thought it was the end of Tom being so consumed with his Charlie Company Swamp Rats. I was so wrong! Now he was on another Charlie Company mission. There were times I became annoyed with Tom for spending so much time on Charlie Company, but I realized how much it meant to him. It was great therapy for Tom – it seemed that he talked much more about his experiences in Vietnam.

We recently attended our third Charlie Company reunion, which was held in Orlando, Florida, May 11-13, 2011. It was a huge success. It's special that the wives and or/girlfriends also attend. The first year we decided the wives would be invited, which turned out to be a great idea. At first we thought if the wives didn't actually want to sit in the meetings, they could go shopping or find other things to do. Little did we know that the meetings would turn out to be so interesting, the wives attended every one! We have a great group of people, and everyone has bonded very well. The first reunion was a little awkward because no one really knew each other, and the men also had to get to know each other all over again. By the third reunion, this had all changed.

I had no idea the effect this senseless war made on so many of the men I met at the reunion. It's so sad to see them sitting at the tables looking at actual pictures from Vietnam and all of a sudden they break down in tears. Or I see them talking to each other and all of a sudden tears roll down their cheeks, and they hug each other. When I witness things like this, I immediately tear up and can't imagine the pain they've gone through and will go through for the rest of their lives. I find it so awful they never received the great receptions and homecomings when they arrived home. Now the soldiers that come home from their tours in Iraq and Afghanistan have huge receptions and parades. Not that I begrudge them in any way, but it was so unfair the way the soldiers were treated when they arrived home from Vietnam! So many times they were called names and things were thrown at them. How awful!

In my mind, their homecoming reception is every year we have the 1/18th Charlie Company reunion. I look at these proud, brave veterans that have such a very special bond between them. I am so happy to be married to one of these veterans who sacrificed it all for our country, and I'm also happy to have been involved getting the 1/18th Charlie Company together after all these years!

January 15, 2009 Letter

from Kathy Buffalo Boy

January 15, 2009
Dear Tom Mercer,
We lost Bob due to congestive heart failure on December 27th. He was staying at our oldest daughter Whitney's house…. He would have enjoyed your reunion and was looking forward to seeing all of his buddies.

We buried Bobby with a full military funeral. The Honor Guard from Bismarck came down along with the various American Legions around this area.

I'm sorry for not introducing myself to you when you first contacted Bobby. I'm his wife, Kathy.
Kathy Buffalo Boy, Wife of Robert Buffalo Boy

January 22, 2009
Dear Kathy,
Words cannot express how saddened and shocked I was to receive your package with the terrible news about Buffalo Boy. I, and our other Army buddies, were really looking forward to seeing him at the

1968. Nov. Platoon, rowdy bunch. Smokey on the left, Buffalo Boy on the right.

reunion. He will be greatly missed, and we will have a moment of prayer at the meeting in his honor. His spirit will be at the reunion with the rest of us warriors from 1966-1968…. I have emailed all of the men that are going to attend the reunion, and everyone has told me to tell you their thoughts and prayers are with you and your family. Mine are as well. Again, I am so saddened by your loss and our loss of a true friend. Even though we haven't seen each other in over 40 years, I know we were in his mind like he was in ours…. Buffalo Boy was a great "point man" during the war. We have an M-14 Club, and I will be mailing his patch to you. We're only giving this patch to "Swamp Rats" who carried M-14s; like the one Buffalo Boy had in his hands from the picture in his obituary.

Take care and I hope you and your family understand Buffalo Boy was a wonderful, brave soldier and will be greatly missed by all of us.

Sincerely,

Tom Mercer

A Passing of Birds

Traci Gardellis (wife of Ken Gardellis)

I did not know my husband, Ken, before his year in country nor did I meet him upon his return home. In fact, I wasn't even born until a good six years after the last helicopter left the embassy rooftop in Saigon.

Living with Ken is like living with the most fun best friend you could imagine. He loves learning, traveling, entertainment, nature, and so many countless other things. We share more interests between us than I do in total with the rest of my friends, and he treats me like a princess. Our relationship is one that has defied the odds of our unusual age gap, and every day brings a strengthening of our bonds. I have learned that "my" Vietnam veteran was a rare find – a proverbial needle in a haystack.

I've also learned a lot about that gritty interior layer dealt with by so many veterans on a daily basis: PTSD. I had learned about the history of the war and the social issues back home, and I have a particular interest in the overall culture of that era. I also had a base knowledge of PTSD. However, until you live with someone who battles the disorder day in and day out, you'll never be able to really get a grasp. It's a bit like a tornado in that we can know in general how it forms and often what triggers it, but we can't always predict its path or the level of destruction it will leave behind.

Sometimes events happen that trigger a reaction that surprises both Ken and me. One such example occurred in May of 2011 at our home in the mountains of western North Carolina. For weeks we had been watching a mother bird, a species known as a phoebe, meticulously building a nest on a little corner shelf we had installed under the overhang of our front entranceway. As with any bird, instinct drove her to find bits and pieces of grass, hair, twigs, and other small building materials to glue together with her own sticky saliva – ultimately producing a quaint villa in which to raise a family. Indeed, one day we lifted up a hand mirror and found five precious eggs nestled in the newly constructed home.

We checked on the eggs every few days and eventually noticed a change in the bird's sitting pattern. Once she vacated the nest for a few minutes, we quickly looked up and verified our suspicion that the babies were hatching. Three of them had completely emerged while two eggs remained. The next day we checked again and found all five of the teeniest, tiniest little baby birds you could ever imagine, all hunkered down together in the base of the nest. We reveled in the births as though they were our own and noted that they occurred just after Earth Day – a fact we thought quite auspicious.

Almost a week passed as I continued my daily classes at the local university, and Ken worked on his personal writings and took care of our cat. We checked on the babies as much as we could without disturbing the mother bird. However, one day as I came home from school, I noticed that the mother bird was not in her nest. Ken asked me if I saw the bird on the way in, and he said that he hadn't seen her since this morning when she had been acting a little strange. He noted that either she or the father bird (both of which were taking care of the babies) had flown and hovered in front of the window near the nest earlier that morning. We both peered out the door's small window, which allowed a close view of the nest and still saw no mother phoebe. Worried, we gathered the hand mirror and went outside. The view in the mirror showed nothing but one solitary baby bird – motionless.

The feelings that came up were a mixture of confusion, sadness, and frustration. We stood under the nest verbalizing as best we could our feelings, which ended up being a combination of "S°°°!" and "I just checked on them this morning!" Even though we could never be sure, we figured the local blue jay had probably taken the babies to feed its own. Eventually, this realization brought out anger in Ken that I rarely see; he began shouting that he wanted to kill the blue jay and that it was a despicable murderer. Ken is one of the most gentle, nature-loving people that I know, so this angry outburst was particularly unusual. I tried to reassure him that the blue jay babies might not have survived without this, but it was to little avail.

Ken's reaction began to go beyond anger and sadness. He cried, and he began to put the blame on himself for not keeping watch. He explained that when the bird had hovered in front of the window earlier in the day, it might have been a cry for help. "Why didn't I listen?" he wondered aloud. Nature could certainly work this way. Maybe the bird was asking for help, but Ken began to take on too much blame because of it. We cried together and tried to stay strong and keep talking through it.

Eventually Ken admitted that his feelings were not merely about this particular incident but also about Vietnam. He equated the role of watching over the safety of the birds to watching over the safety of the men in his squad. These issues of guilt, blame, anger, sadness, and death transferred from present to past and back again. It took a while to work through these feelings that the passing of the babies brought up in Ken. I know he took some time to write about the incident, but the pain was still present.

A day after the death, I gathered up some incense, stones, and prayers in order to have a ceremony for the babies. We carefully lowered the silent remaining baby from its nest and dug a small hole in our beautiful hillside garden. We performed a small ritual smelling of sage and wet with our tears. In a way it put to rest not only the bird and its siblings in spirit but also the guilt and memories it brought up as well.

I couldn't have expected that the passing of these tiny birds would bring up a connection to Vietnam, but over the years I have learned to expect the unexpected. Each case of PTSD is completely different, and there are few words of advice I could offer to other wives because of it. The things I know for sure, though, are that some combination of love, patience, laughter, and fun has eased the severity and frequency of episodes.

I also know PTSD has been at times almost debilitating for Ken and so many others. Ken received his 100% disability so he could leave the stress of work. In a way I think he wouldn't be the same unique and wonderful husband he is today if not for his particular experiences of Vietnam and its aftermath. However one explains the success of our relationship, one thing is absolutely true: he is and will always be my hero.

A Soldier's Musings and Laments

Patrick McLaughlin

At the beginning and end of the day, we went to Vietnam alone, and we came back alone. One day we were with the guys we grew up with in that life-changing tour of duty, and the next separated, mostly forever. Back in the world we went about the business of the rest of our lives. For some the transition was easier than others. War does that to young men – old men too.

Like feathers blown by a gust of wind we went in all directions never to be rejoined. Charlie Company was no exception. Yet, we have come together although it took over 40 years. Our memories have dimmed, hair receded, morphed to grey, and some belts enlarged. But the unique bond among men who paid their generations' dues remains, and this is special.

At the first reunion, May of 2009, Nam plus 42 years, Lima Platoon from my era was back together. There was O'Connor,

Di An. 1967. Lima Platoon undergoing weapons inspection.

Coleman, Duncan, Mercer, Gardellis, Smith, Norris, Myrick, Cone, Gilbert, Estus, and Smart. We learned that Beal had passed away, but his son came. Bill Annan, Charlie Six, was present, and we had a speaker phone call with Dogface Six, General Cavazos. One Charlie Company soldier said, "Sir, I've never had the chance to congratulate you on making general."

Cavazos replied, "Why, you all made me a general." Classic Dogface Six, a real leader of men.

May 23, 1985. The National Vietnam Veterans Network held its first annual awards dinner, a black-tie affair, at the Sheraton Washington Hotel in the nation's capital. Ten Vietnam veterans were honored, and there were several distinguished guests of honor at the event including a recipient of the Congressional Medal of Honor. Following the dinner, Christine and I were invited by Marge and Mike Gavin to the suite of Bob Hope. Marge Gavin was Bob Hope's niece, and Mike and I later became law partners.

General William Westmoreland was present in Hope's suite, both distinguished guests of the awards event. It was a lovely night so the four men decided to take a walk. Mike and Bob Hope walked ahead and General Westmoreland and me behind. The general, commander of all U.S. forces in Vietnam from 1964-68, turned to me and said, "Tell me again what unit you served with in Vietnam."

I replied, "Charlie Company, 1st Battalion, 18th Infantry, 1st Infantry Division."

Westmoreland asked, "Who was your battalion commander?"

"Lieutenant Colonel Richard E. Cavazos."

"Dick Cavazos," Westmoreland stated, "was one of the finest battalion commanders I saw in Vietnam."

Just as we Dogface soldiers will always remember Dogface Six, he, likewise, remembers his soldiers. In early October of 1984 I was appointed by the judges of the U.S. district court for the Northern District of Ohio as the United States Attorney. While serving in that post as the chief federal law enforcement official in northern Ohio, I was under consideration by President Ronald Reagan for his appointment, with the consent of the Senate, as U.S. Attorney. The Cleveland Plain Dealer did a story that ran on December 30, 1984 under the headline "U.S. Attorney Nominee Likes to 'Walk the Point.'" The PD reporter contacted General Cavazos and quoted him in the article. The article reads that "retired four-star General Richard Cavazos remembers 'Mac' as the man who volunteered to walk the point more than any of the other 1,000 members of the battalion. It was not bravado, but McLaughlin's high sense of purpose and responsibility, said Cavazos from his Texas home."

"We have a saying in the infantry," Cavazos stated. "The best infantryman of all infantrymen is he who walks the point. He's just the guy who is apt to get cut down." The article states that Cavazos described me as "one of the top combat soldiers he ever met in his 34-year Army career, a type of man he rarely sees." The article concludes the quotes from Dogface Six, "I thought the world of him. He could have my wallet anytime he wants it."

Well, General Cavazos, you can have mine anytime you want it and the wallets of a bunch of other Dogface soldiers who believe as I do that we served with "one of the finest battalion commanders" of the Vietnam War.

Last Day in the Field. Over the years I have searched for Ray (Porky) Etherton to no avail. We would all rejoice in seeing him again. Etherton and I went through training together and ended up in the same unit in Nam. We left the field together on Christmas Eve 1967.

I walked the point on a platoon-sized patrol off Thunder Road between Lai Khe and Quan Loi, as a parting gift to my guys on that Christmas Eve. When Bill Annan learned that I was walking point, he came straight to Lima's position. The conversation went something like this.

"Sergeant Mac, what are you doing?"

"What do you mean, Sir?"

"I am informed that you intend to walk the point today."

"I'm taking the point out."

"No one walks point on their last day in the field."

"That's true, but I'm walking it today."

Captain Annan looked me squarely in the eyes and said, "Sergeant Mac, I have a mind to order that you not walk the point today. In fact, you don't have to go out on the patrol."

"Captain Annan," I replied, "I would consider it a personal favor to me that you not issue that order. I know what I'm doing."

Annan glared at me, shook his head slightly, and then turned to walk off. "Sir, appreciate it." Charlie Six turned, and I might have detected a crack of a smile as he glanced at Lima Six, Lieutenant Emmett Smart, and said, "Carry on." Bill Annan and Charlie Company forged an unbreakable bond at Loc Ninh. He knew that he could count on us and us on him.

Best Nam Advice. Actually, Annan gave me the best advice I received during my tour of duty. It was after Thanksgiving, and we were sent back to Di An to catch our breath and kick back for a couple days. Arriving at the Charlie Company area, the NCOIC had filled a couple large containers with ice and tossed in beer and pop. We all were filthy as it had been more than a week since a change of jungle fatigues and longer since our last shower. At that point one is no longer offensive to oneself or others similarly situated, only to those who have recently showered. I tossed my gear in the hooch and headed straight to the cold drinks. I fished out a cold Coke and was enjoying how magnificent it felt going down when someone asks, "Are you Staff Sergeant McLaughlin?"

"I am." Turning, I am face-to-face with a captain, clean jungle fatigues and shined jungle boots. He wears jump wings, a Ranger patch, and Special Forces designations. This is a gung-ho officer, but nothing I see tells me whether he has spent any time in combat. They don't teach the real thing at any training course or school. The captain then informs me that he has the authority of Lieutenant General so-and-so and is offering me the E-6 slot on his six-man special ops team that is directed to establish an CIDG unit and compound in the Delta. My job would be to lay in the defensive perimeter and train the unit on ambush patrols. Surprised that the captain has come to me, I comment that I am a straight-leg grunt, so why me? He stated that I was highly thought of at division headquarters and the recommendation was made from there. When would I have to decide?

"Now," he said, "you would grab your gear and come with me." At this point I've got eleven plus months in Nam, most in the boonies, and the thought of leaving my guys for a group of strangers – mostly Vietnamese – gave me pause, which I conveyed. The captain upped the ante, "By the way, Staff Sergeant, I have the authority of Lieutenant General so-and-so to order you to accompany me and leave now." "Let me talk it over with my CO." I went to Annan who knew why the captain had come since he had paid a courtesy call on Charlie Six before approaching me. After over viewing the captain's "offer" I asked, "What do you think?"

Bill Annan said, "Mac, you've had a long year. You have been lucky. Go home." Damn good advice. Realizing that the captain needed me to agree to extend my tour to accomplish his mission, I told him that I was flattered that he sought me out but declined his offer. He countered by reminding me that he had the authority to order me to accompany him and depart the Charlie Company compound within the hour. He likely wanted me to clean-up before going anywhere with him.

"Captain," I said, "you can take me with you, but when my year is up I'm going home on leave." There was some negotiation, but once he determined that my resolve was firm, he extended his hand. "Good luck with your CIDG unit," I said.

"Thanks," he replied, "I'm going to need it."

In early January 1968 I departed Vietnam after a long and lucky year of combat. My last active duty months in the U.S. Army were served at Fort Huachuca, Arizona, as a platoon sergeant in an AIT unit.

When that golden bird lifted off in early January of 1968, and a collective, spontaneous cheer erupted, it did not occur to me that I would ever return to Nam. But I did return, in 2005, with my wife of 35 years, and spent ten days in country. I didn't go back to any of the places frequented on the first go around – names like Loc Ninh, Di An, Lai Khe, Quan Loi, Phouc Vinh, Song Be, Phu Loi, Highway 13 (Thunder Road) and, more generically, War Zones C and D.

September 1985. Bob O'Brien and I stayed in close touch after our service together. We continued to correspond after I returned from Nam and O'Be completed his third-consecutive tour of duty. When he returned stateside, military obligation done, O'Be came to Cleveland and met my family and bride to be, the ever-engaging Christine. O'Be settled in Florida and furthered his education at St. Leo College, meeting his future wife, Janie. They married and had a son, Patrick Michael.

Like some Vietnam veterans, Bob O'Brien never really made it back from that war. He sought treatment at the Veterans Administration for his PTSD issues. O'Be was treated by the VA as Vietnam vets were treated by the country – if not as outcasts then certainly as unwanted step-children. The VA botched his medications and never actually got his treatment right. O'Be consistently complained to me about the medications prescribed by and insensitivity shown to him by certain professionals. He grew increasingly depressed.

One September evening around 10:30, maybe later, O'Be called to talk. This was not unusual, although the call was later than customary. Bob was down, more so than usual, and he was upset about going in the next day for a medical procedure at the VA. We spoke for 30 minutes or more. I did my best, I hoped, to perk him up and remind him of all the blessings he had in his life. When we ended, O'Be said, "Goodbye, Mac."

As the call ended, I was unsettled but couldn't put my finger on why this was so. About 15 minutes passed and the phone rang. It was Janie. She was hysterical. Bob had gone out in the backyard after we talked, smoked his last cigarette, and ended his life. The details remain too painful to reflect upon.

That night I kept replaying over and over our last conversation when I understood what had unsettled me. In all our time together Bob had always said, "So long, Mac" when we departed in person or by phone. My friend had called this time to say, "Goodbye, Mac."

In 1985, the men of Lima Platoon were feathers blown throughout the country. We had not reconnected. Only John O'Connor and I attended O'Be's funeral, but if we had been able to notify the others, many would have come to pay tribute to a fallen brother and offer condolences to his wife and son.

I gave the eulogy on that September 22nd day of 1985, and I hope that, in some small way, it was worthy.

IN MEMORY OF ROBERT D. O'BRIEN (1938-1985)

Janie and Patrick Michael O'Brien, Bob's brother Bill O'Brien, family and friends of my dear friend Robert O'Brien.

Patrick, when you have been around this world for as long as I and other adults here this evening have, you will come to understand that over the course of our lifetime a mere handful of individuals with whom we come in contact have an enormous impact on our life. These few individuals are truly unique. They are exceptional persons, men or women, who demonstrate such strength, compassion, integrity, and courage that they leave an indelible imprint on our lives. Your father, Bob O'Brien, left such an imprint on my life.

I and John O'Connor first met your dad in Vietnam in January, 1967. We all served together with Charlie Company, 1st Battalion 18th Infantry, 1st Infantry Division (the Big Red One). Your dad and I became squad leaders in the same platoon – Lima Platoon.

Your father was an exceptionally brave soldier. He won the Silver Star for gallantry in action at Loc Ninh, Republic of Vietnam, on October 29, 1967. He was a great leader of men in combat because he possessed all of the attributes of leadership. He was a leader of men and not a follower. Most importantly, Patrick, your dad possessed the one true mark of a leader of men in combat. He never asked anyone serving under him in combat to do that which he would not, or indeed did not, himself do.

Bob O'Brien served with the Army in Vietnam for three years. A three- year tour of service was extremely rare. Most of us served only one year in Vietnam. Your father volunteered for those extra years because of his commitment to duty and his love for his country. But the war invoked its price. We who served in that war know that not all of the casualties occurred on the battlefield. Bob O'Brien carried the scars of that war long after hostilities ended.

Your dad was an American hero. Today, in my mind and in the mind of Johnny O'Connor, one more name has been added to the black granite wall in Washington, D.C. – The Vietnam Veterans Memorial. There are now 58,023 names engraved in history on that memorial, a most eloquent tribute to the Vietnam Veteran. Patrick, you should never forget that your father was a hero.

You should never forget that your father loved you and your mother more than anything on this earth. I know, because he told me so many times. And your mother loved him and cared for him when he most needed it. To me, that depth of love is a tribute to mankind.

I remember a poem I studied in high school entitled "Lament," by Edna St. Vincent Millay. In that poem, a mother tries to explain to a young boy and girl the death of their father. She concludes by saying:

> "Life must go on,
> and the dead be forgotten;
> Life must go on,
> Though good men die;
>
> ∞
>
> Life must go on;
> I forget just why."

At times like this, it is hard to remember why life must go on. But it must and it should.

As life goes on, there are some things we do need to forget – and it is important that we can forget. But other things must be *remembered. Patrick, I think it is important that you always remember what I have said tonight about your father.*

Almighty God, Lord of us all, have mercy on the soul of Robert O'Brien. He was a soldier, a good man, a caring and sensitive man, and a loving husband and father. O'Be, my friend, may you rest in peace. Amen.

Welcome Home

Caroline Cavazos

The tragic attacks on 9/11 with the loss of so many innocent lives reminded Americans that we will always have to be prepared to defend ourselves. Evil exists in the world in spite of the efforts of good people to deter it.

World Wars I and II were "popular" wars. Men signed up to be part of the adventure. Songs were written and sung with enthusiasm. The world appeared to be a better place when those wars came to an end. Korea was different. There was not a lot of enthusiasm for another war so soon after WWII. But men in college on the G.I. Bill dropped out of classes and joined units to hold the line and preserve a free South Korea. Again, American soldiers did their bit.

We fought next to preserve South Vietnam – a very controversial war. Again, American soldiers did their duty to carry out U.S. policy – whether or not they understood it. No popular songs were produced. Anti-war was the general feeling at home. Few returning servicemen received a "thank you" or "well done, soldier" – a sad comment on the U.S. in the 1960s and 1970s. But that does not change the fact that you served bravely and with honor when your country asked. You did your duty; it was your country-

January 1968. John Claire, Billy Mack.

men who failed you!

Once again we face threats from afar. American troops are fighting and dying to keep the enemy from our shores. Be proud to have done your part – and done it well – when it was your turn to keep America free. Thank you for your service, Dogface Charlie Company soldiers, and Welcome Home!

Saddle Up and Move Out

Tom Mercer

There's no easy way to end the stories about Charlie Company of the 1st Battalion, 18th Infantry. They could go on and on, like the brotherhood that was formed with the guys of Charlie Company and has never ended to this day. The firefights and battles we fought taught us how to become men. Over the past 43 years, even though we weren't together, we had to grow up again. Now we are husbands, fathers, and grandfathers. Our whole life has passed before us, but the brotherhood we shared years ago is still going strong. Even though we don't remember all the names, the faces will always be etched in our memories forever. The men who were killed in battle, and the men who were wounded in battle, will always be stored in my memory bank, so I can bring them out whenever I want. I am honored to have served with the Big Red One, Charlie Company, Dogface, Swamp Rats, from 1967-1968. God bless the men from Charlie Company.

Appendix 1

1-18th Comrades Lost in Vietnam*

	Last Name	First Name	Middle Initial	Rank	BN	Company	Date KIA
1	Gaspar	John	J.	2LT	1/18	B	18-Jul-65
2	Bearwald	Orlando	Orrin	PFC	1/18	HHC	8-Sep-65
3	Bailey	Raymond		SP4	1/18	B	6-Oct-65
4	Holbrook	James	Newton	SFC	1/18	B	6-Oct-65
5	Holton	Leon	G.	1LT	1/18	A	8-Oct-65
6	Lynch	Richard	E.	PFC	1/18	A	8-Oct-65
7	Mason	Bobby	G.	SGT	1/18	A	8-Oct-65
8	Roden	George	Columbus	SP4	1/18	A	8-Oct-65
9	Ron	Grijalba	Humberto	SP4	1/18	B	8-Oct-65
10	Breeding	Wayne	Peter Earl	SGT	1/18	A	9-Oct-65
11	Liptock	Michael		PFC	1/18	C	13-Oct-65
12	Dickens	David	R.	1SG	1/18	B	1-Nov-65
13	Eklund	Paul	Herbert	2LT	1/18	A	19-Nov-65
14	Jensen	Arlin	Roger	SP4	1/18	A	19-Nov-65
15	Reeder	David	Lee	SGT	1/18	A	19-Nov-65
16	Chandler	Connie		SGT	1/18	B	20-Nov-65
17	Emerling	John	P.	PFC	1/18	B	20-Nov-65
18	Sargent	Kenneth	E.	SGT	1/18	B	20-Nov-65
19	Trypu	Frank	Donald	SGT	1/18	A	20-Nov-65
20	Wilson	Billy	Joe	SP4	1/18	A	20-Nov-65
21	Hammond	Kenneth	Joe	SP4	1/18	B	22-Nov-65
22	Smith	Cleo		CPL	1/18	A	24-Nov-65
23	Thompson	Sammy	Lee	PFC	1/18	A	24-Nov-65
24	Keckler	Robert	L.	SP4	1/18	B	10-Dec-65
25	Thompson	George	R	SGT	1/18	B	11-Dec-65
26	Horne	Kenneth	Raymond	PFC	1/18	C	12-Feb-66
27	Hirsch	Marshall	Raymond	2LT	1/18	C	21-Feb-66
28	Brown	James	Truly	PFC	1/18	B	26-Feb-66
29	Winningham	Clifton		PSGT	1/18	B	5-Mar-66
30	Lankford	Henry	D	PFC	1/18	B	5-Mar-66
31	Nutt	Richard	E	PFC	1/18	B	5-Mar-66
32	Heiser	Duane	Kenneth	PFC	1/18	A	15-Mar-66

33	Misheikis, Jr.	Theodore	N.	SP4	1/18	C	26-Mar-66
34	Martinez	Isidro		SP4	1/18	B	30-Mar-66
35	Faulkner	Michael	Lee	PFC	1/18	B	8-May-66
36	Johnson	Arthur	Louis	PFC	1/18	B	8-May-66
37	Pierce	William	Wesley	SP4	1/18	B	14-May-66
38	Bermea	Victor	C.	SGT	1/18	A	29-May-66
39	Harrigan	Lawrence	Colburn	PFC	1/18	A	8-Jul-66
40	Newkirk	Thomas	Clifton	PFC	1/18	C	10-Sep-66
41	Willis	M.	L.	SGT	1/18	C	12-Sep-66
42	Calloway	Larry	James	SGT	1/18	C (1st Sqd, 1st Pltn)	25-Sep-66
43	Bailey	Larry	William	SP4	1/18	C	2-Oct-66
44	Byam	Michael	Leroy	PFC	1/18	C	2-Oct-66
45	Haddock, Jr.	Louis	Edward	PFC	1/18	C	2-Oct-66
46	Pittman	James	Sherwin	PFC	1/18	C	15-Oct-66
47	Gonzalez	Rodolfo (Rudy)	Guadalupe	SP4	1/18	C (1st Sqd, 1st Pltn)	11-Nov-66
48	Bass, Jr.	Duncan	Edward	SGT	1/18	HHC	18-Nov-66
49	Gilbert	James	Silas	SP4	1/18	HHC	18-Nov-66
50	Victoria	Frederick	Pearce	1LT	1/18	HHC(Recon)	18-Nov-66
51	Cemelli	Salvatore	Peter	PFC	1/18	B	8-Dec-66
52	Harbin	Gary	Lee	PFC	1/18	B	8-Dec-66
53	Nadolski	Robert		PFC	1/18	C	1-Jan-67
54	Arthur	James	Raymond	SP4	1/18	HHC	9-Jan-67
55	Chatmon	Nathen	Eugene	SP4	1/18	HHC	9-Jan-67
56	Flanagan	Tom		SGT	1/18	HHC	9-Jan-67
57	Gould	Caryle	Leroy	SP4	1/18	HHC	9-Jan-67
58	Merrill	David	B.	PFC	1/18	HQ	9-Jan-67
59	Quinn	Bobby	Joe	PFC	1/18	HHC	9-Jan-67
60	Schollard	John	Andrew	PFC	1/18	HHC	9-Jan-67
61	Bullock	Nathaniel		PFC	1/18	D	15-Jan-67
62	Miles	David	Lee	PFC	1/18	B	15-Jan-67
63	Otte	Kenneth	Michael	PFC	1/18	B	16-Jan-67
64	Torres	Victor	Luis	PVT	1/18	B	17-Jan-67
65	Narum	Thomas	Leroy	CPL	1/18	A	18-Jan-67
66	Contreras	Pablo	Guereca	SP4	1/18	B	20-Jan-67
67	Davis	Billy	Sylvester	SSGT	1/18	B	20-Jan-67
68	Davenport, Jr.	David	D.	PFC	1/18	C	5-Feb-67
69	Mezzles	Tommy		SP4	1/18	HQ	28-Feb-67
70	Johnson	Russell	Carl	SP4	1/18	C	4-Mar-67
71	Mills	Donald		SSGT	1/18	A	25-Mar-67
72	Matthews	Willis	Alanzo	PFC	1/18	A	4-Apr-67

73	Vanbuskerk	Harold	Dennis	PFC	1/18	A	9-Apr-67
74	Miller	Charles		PFC	1/18	C	17-Apr-67
75	Rainey	William	George	PFC	1/18	HHC	6-May-67
76	Prys	Robert	William	PFC	1/18	A	29-May-67
77	Cossa, Jr.	William	Edward	SP4	1/18	C	31-May-67
78	Williamson	Paul	Douglas	SP4	1/18	HHC	31-May-67
79	Davis	Willie	Louis	PFC	1/18	HHC	14-Jul-67
80	Wagner	Roy	Carl	SP4	1/18	A	2-Oct-67
81	Dossett	James	Edwin	PFC	1/18	C	5-Oct-67
82	Dingle	Earl		SP4	1/18	B	6-Oct-67
83	Oestreicher	Paul	Anthony	SP4	1/18	B	6-Oct-67
84	Dresher, Jr.	Harry	Everett	SP4	1/18	B	11-Oct-67
85	Cruz	Sam		SP4	1/18	???	12-Oct-67
86	Gentry, Jr.	Charles	Edward	PFC	1/18	C	29-Oct-67
87	Amos	Joe		PSGT	1/18	A	30-Oct-67
88	Hanson	Kenneth	Gregory	SGT	1/18	A	30-Oct-67
89	Kenter	Michael	William	SGT	1/18	A	30-Oct-67
90	Tomlinson, Sr.	James	Howard	SFC	1/18	D	30-Oct-67
91	May	John	Albert	SSGT	1/18	C	2-Nov-67
92	Chambers	Robert	Stanley	PFC	1/18	B	24-Nov-67
93	Lucas	John	Willie	SP4	1/18	B	24-Nov-67
94	Mackey	Vertis	L.	SP4	1/18	B	24-Nov-67
95	Chesnut	Gerry	George	SSGT	1/18	D	2-Dec-67
96	Boger	Rhinehart		SP4	1/18	A	10-Dec-67
97	De Waal	Howard	Jacob	SP4	1/18	A	11-Dec-67
98	Myers	George	Laxley	SGT	1/18	B	12-Dec-67
99	Ruminski, Jr.	Philip	Edward	SP4	1/18	B	7-Jan-68
100	Garcia	Joseph	Andrew	SP4	1/18	A	31-Jan-68
101	Alberts	Roger	Duane	PFC	1/18	HHC	5-Feb-68
102	Leonard	Sidney	Lamar	CPT	1/18	HHC	7-Feb-68
103	Bryan	Dan	E.	PFC	1/18	D	8-Feb-68
104	Thompson, Jr.	Onnie		SP4	1/18	C	8-Feb-68
105	Granados	Richard		SP4	1/18	C	5-Mar-68
106	Ambrose	Gregory	Francis	PFC	1/18	B	15-Mar-68
107	Taggart	Larry	Joel	PFC	1/18	C	15-Mar-68
108	Shutters	Patrick	Alanzo	1LT	1/18	B	15-Mar-68
109	Thompson	Charles	Clair	SP4	1/18	C	11-Apr-68
110	Maroscher	Albert	George	MAJ	1/18	HHC	15-Apr-68
111	Deeter	Avid	Kim	PFC	1/18	C	19-Apr-68
112	Mello, Jr.	Edward	Thomas	1LT	1/18	C	19-Apr-68
113	Thompson	John	Bryan	SGT	1/18	A	20-Apr-68
114	Okemah	John		SFC	1/18	D	28-Apr-68

115	Bonifant	Samuel	Harold	SSGT	1/18	B	4-May-68
116	Johnson	William	William	SGT	1/18	D	4-May-68
117	Klein	Gerald	Dean	SGT	1/18	D	4-May-68
118	Mills	Richard	Thomas	SP4	1/18	D	4-May-68
119	Price	Robert	Glenn	2LT	1/18	D	4-May-68
120	Torres	Anthony	Wilfred	PSGT	1/18	A	4-May-68
121	Hinnant	Benjamin	Lowell	SFC	1/18	A	6-May-68
122	Moore	Terry	Lee	PFC	1/18	B	6-May-68
123	Cooper	James	Ennis	SP4	1/18	D	8-May-68
124	Lee	James	Richard	PFC	1/18	B	12-May-68
125	Seltzer	Jackie	Ralph	SGT	1/18	B	12-May-68
126	Gibson	David		SP4	1/18	B	18-May-68
127	Veihl	John		PFC	1/18	D	19-May-68
128	Carter, Jr.	Hamp		SGT	1/18	D	3-Jun-68
129	Hall	Chauncey	Ike	SSGT	1/18	D	3-Jun-68
130	Oakes	Jack	Wayne	PFC	1/18	C	17-Jun-68
131	Denton	David	Andrew	SP4	1/18	A	23-Jun-68
132	Shields	David	Thomas	PFC	1/18	B	29-Jun-68
133	Arenas, Jr.	Manual	V.	PVT	1/18	D	9-Jul-68
134	Romero	Robert	Anthony	SP4	1/18	B	27-Jul-68
135	Tafoya	Joseph	Ernest	PFC	1/18	C	28-Aug-68
136	Adcock	Billy	Anthony	PFC	1/18	A	4-Oct-68
137	Clark	Douglas	Mark	SP4	1/18	E	4-Oct-68
138	Miller	Hubert	Ayne	SP4	1/18	A	4-Oct-68
139	Smith	Paul	Leslie	PFC	1/18	C	5-Oct-68
140	Bartkowski	Gregory	Joseph	PFC	1/18	C	30-Oct-68
141	Horsley	Richard	Wayne	PFC	1/18	C	30-Oct-68
142	Hellenbrand	David	Peter	SP4	1/18	B	1-Dec-68
143	Terry	Philip	Allen	SGT	1/18	A	8-Dec-68
144	Derrick	Randy	Wayne	SP4	1/18	A	13-Dec-68
145	Fray	Earl	Richard	SGT	1/18	A	13-Dec-68
146	Russek	John	Joseph	SGT	1/18	A	13-Dec-68
147	Walden, Jr.	Marion	Frank	PFC	1/18	A	13-Dec-68
148	Puentes	Miguel	Angel	SP4	1/18	HHC (Recon)	26-Jan-69
149	Winkel	William	Daniel	PFC	1/18	HVY PLT	31-Jan-69
150	Freppon	John	Dennis	SSGT	1/18	E	2-Feb-69
151	Morrison	James	John	SP4	1/18	E	2-Feb-69
152	Robinson	James	Marcus	SGT	1/18	B	2-Feb-69
153	Coyle	Gary	Joseph	PFC	1/28	D	7-Feb-69
154	Boyd	Charles		SP4	1/18	HHC	15-Feb-69
155	Monish	Ronald	Anthony	SP4	1/18	D	15-Feb-69
156	Fitzgerald	Manfred	Willy	SGT	1/18	D	20-Feb-69

157	Walters	William	Owen	SGT	1/18	D	3-Mar-69
158	Kangro	Lauri		SP4	1/18	B	4-Mar-69
159	Frericks	Louis	Wayne	1LT	1/18	C	12-Mar-69
160	Young	Jerry	Owen	PFC	1/18	C	12-Mar-69
161	Beasley	Edward	Russell	PFC	1/18	D	14-Mar-69
162	McCarthy	Brian	Francis	SP4	1/18	D	15-Mar-69
163	Parham	John	William	SGT	1/18	C	18-Mar-69
164	Davison	Jackie	Lee	PFC	1/18	C	19-Mar-69
165	Nolley	Lee	Roy	PFC	1/18	???	5-Apr-69
166	Owen, Jr.	William	Lee	2LT	1/18	HHC	11-Apr-69
167	Zapolski	Lawrence	Edward	SP4	1/18	B	11-Apr-69
168	Parker	Danny	Lynn	CPL	1/18	D	6-Jun-69
169	Rosario-	Soto, Jr.	Ernesto	SP4	1/18	HHC	6-Jun-69
170	Shelton, Jr.	Dallas	C.	SP4	1/18	B	20-Jun-69
171	Poff	Jerry	Wayne	SP4	1/18	C	3-Jul-69
172	Sikes	Charles	Michael	SP5	1/18	HHC	28-Jul-69
173	Walker	Leslie	Elroy	SP4	1/18	E	28-Jul-69
174	Cooper	Robert	Wesley	SP4	1/18	D	8-Aug-69
175	Shipley	Ronald	Eugene	SSGT	1/18	D	16-Aug-69
176	Wood	Donald	Fred	1LT	1/18	E	19-Aug-69
177	Spilker	James	Dennis	SGT	1/18	A	20-Aug-69
178	Custode	Ralph		SGT	1/18	B	24-Sep-69
179	Tonon	James	Anthony	SP4	1/18	D	27-Sep-69
180	Mosher	Alex	Roy	SP4	1/18	C	8-Oct-69
181	Marrington	Craig	Thomas	SP4	1/18	D	20-Oct-69
182	Beer	Merlin	Gail	SP4	1/18	B	11-Nov-69
183	Kelley	Harvey	Paul	CPT	1/18	A	20-Nov-69
184	Atwell, Jr.	Donald	William	PFC	1/18	B	27-Dec-69
185	Jenneges	Ronald	Arthur	SGT	1/18	B	28-Dec-69
186	Bulla, Jr.	Robert	Franklin	SSGT	1/18	D	22-Jan-70
187	Neeley	Lowrenzo		CPL	1/18	A	22-Jan-70
188	Nunley, Jr.	Walter	William	SGT	1/18	A	22-Jan-70
189	Cobb	Roy	William	SFC	1/18	B	13-Feb-70
190	Smeltzer, III	Charles	E.	SGT	1/18	B	13-Feb-70
191	Giles	Claude	Vernor	SGT	1/18	A	16-Feb-70

* Special thanks to Larry Van Kuran for compiling this list. It is not official, and some errors may exist, but it is a work in progress. If you know of any corrections that need to be made, please contact the First Division Museum so that updates can be made to future printings.

Appendix 2

Table of Organization & Equipment Explanation in Vietnam
Paul Douglas Goddard Jr.

W hile the 1st Infantry Division was at full strength, rifle companies were always understrength. Platoons routinely fielded 24 to 30 men rather than 44 including headquarters of platoon leader, platoon sergeant, medic, and radio telephone operator (RTO). Squads would have seven or eight men rather than ten. The weapons squad normally was two machine gun teams assigned to different rifle squads. At times, Charlie Company platoons ran two squads of 12 to 14 men, which included the M-60 gun team of four.

To have a career soldier staff sergeant leading a squad was the exception rather than the rule. Most squad leaders were sharp soldiers that moved up the ranks during their RVN service. On rare occasion, these soldiers were taking charge as a private first class and later achieving the rank of sergeant. Platoons were usually led by lieutenants, although platoon sergeants were occasionally in charge of platoons. Typically, career noncommissioned officers were platoon sergeants, but by mid-1968, staff sergeant graduates of the Infantry Non Commissioned Officer School were being assigned as platoon sergeants. There were times that a 23- or 24- year- old lieutenant would be the oldest person in a platoon.

While one third of the United States Army was in the RVN during 1968-69, one half of the captains were in country. For an extended period in 1968, two of the four rifle companies and headquarters company of the 1st Battalion 18th Infantry were commanded by first lieutenants rather than captains. Lieutenants routinely filled in as company commander on a temporary basis. On many occasions the three rifle platoons and weapons platoon of Charlie Company accomplished its mission with barely 100 combat ready men in the field. Under the adversity of low numbers and years of service, the best-trained, most-educated, and youngest US Army did what they were asked to do and did it exceptionally well.